SOMETHING
in the
AIR

A FOUR-SEASON GUIDE
TO OUR HEALTH
AND THE WEATHER

ANTHONY R. WOOD

Prometheus Books

Essex, Connecticut

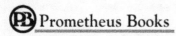 Prometheus Books

An imprint of The Globe Pequot Publishing Group, Inc.
64 South Main Street
Essex, CT 06426
www.globepequot.com

Distributed by NATIONAL BOOK NETWORK

British Library Cataloguing in Publication Information Available

Library of Congress Cataloging-in-Publication Data Available

ISBN 9781633889965 (paper) | ISBN 9781633889972 (epub)

To Sean F. and William K. Wood

CONTENTS

PREFACE

Along with suffering, hundreds of millions throughout the world who have endured the pains of arthritis and migraines have at least one thing in common: They insist that weather precipitates and worsens their suffering.

Some of the best minds, those who have attempted to understand and remedy these and other intricate medical conditions, agree that their patients have powerful cases.

The human body and the atmosphere meet at one of the most magical and complex intersections in the known universe. Understanding that dynamic and often mysterious relationship has taken on new urgency with accelerating climate change. In addition to migraines and arthritis, credible research has linked weather and its ever-changing behavior to everything from suicides to sexual activity, from purchasing decisions to diets, from melancholy to irrational exuberance, from insomnia to mental acuity, from mood changes to the behavior of the stock market.

Humans historically have exhibited the ability to adapt to changes in climate, sometimes on short notice. Think of Mexican, African, Caribbean, and Indian immigrants who populate Northeastern U.S. cities. Or seventeenth-century European settlers initially shocked by the harsh winters in the colonies.

But how might rising temperatures affect common seasonal ailments, such as colds and influenza, and the more exotic disease vectors? And how might heat waves intensify on the urban heat islands that already have become summer hot plates due to heat-retaining buildings and paved surfaces?

What follows is a conversational, historical, clinical, and factual tour of the science of *biometeorology*—technically, that's the study of the interactions between the atmosphere and living organisms—and in this book, the discussion is limited to human beings. I address what is known about weather–health interactions, separating reality from fantasy and folklore. I have endeavored to avoid sensationalism: Climate change is real, and people are scared enough.

As researchers cautioned in a National Academies of Science publication, *Under the Weather* (which I highly recommend), epidemiology, climatology, and meteorology are very different disciplines, each with much to learn from the others: "Epidemiological data from many parts of the world are highly limited or nonexistent. . . . In turn, epidemiologists and other health professionals need to understand the considerable uncertainties associated with many aspects of climate forecasting."[1]

Science is, indeed, science, and it is layered with complexities, as is biometeorology. And a day when we see a journal paper that ends with "No more research is needed" will never come.

Once an arcane discipline, biometeorology's revived popularity is evident in the daily "health" forecasts for arthritis, migraines, and allergies that appear on commercial weather sites. But *caveat emptor*, those are the works of Dr. Algorithm.

Through the decades, adventurous and visionary medical researchers and practitioners have done the hard and often frustrating work of seeking the answers to atmosphere–body riddles. I think of the international arthritis expert, whose work is discussed in chapter 2, who consigned those under his care to an experimental chamber worthy of science fiction to prove that weather changes both incite and intensify pain. For his trouble, he endured considerable frustration himself.

This is a plain-language lay treatise, and I have tried to eschew technical medical jargon where possible. This has required some degrees of simplification. For example, in the chapter on vitamin D, I dispensed with the D_2, D_3 distinction. That's acceptable, in the view of Glenville Jones of Queen's University in Canada, who observed that "these two forms have roughly equal potencies, similar metabolic patterns and identical effects in the body."[2]

Much of the reporting is drawn from personal experiences. I confess to being somewhat of an "atmo-chondriac," ever-conscious of the effects of the weather on my mental and physical health.

However, we journalists have our own diluted version of a Hippocratic Oath. In that spirit I have done my best to transcend my own biases and attempt to present objectively what is and isn't known about the body–atmosphere relationship—inescapable even in our managed indoor environments—and why researching this relationship has been so problematic. The literature on the subject is voluminous, and ranges from embracing to dismissive—sometimes both simultaneously. Studies have shown almost everything.

The material in the book is arranged by seasons.

Each section begins with a prelude summarizing the health aspects most commonly associated with that season, although several of the conditions addressed aren't confined to one season. Pollen allergies begin at the very end of the astronomical winter and continue into the beginning of fall. They are discussed in the spring section because in the temperate zone, that embraces the peak periods of two of the big three pollen allergens—trees and grass. Ragweed, the third leg of the pollen trilogy, doesn't ripen until late summer. In any event, maladies have about as much respect for our calendars as the atmosphere does.

Each chapter ends with short summaries of strategies and remedies for mitigating the effects of the various conditions as recommended by the likes of the Centers for Disease Control and Prevention (CDC) and other medical professionals and organizations, including the American Academy of Allergy, Asthma and Immunology and the American Migraine Foundation.

While I have endeavored to separate the legitimate from the unsupportable, it is clear that ignoring the atmosphere in matters of health is as much folly as ignoring the importance of diet and exercise in an intelligent wellness program.

INTRODUCTION

BIOMETEOROLOGY, THE FIRST 4,500 YEARS

- Physicians, patients, and medical researchers who hold that weather and climate significantly affect human health have strong support from the "father of medicine."

- The body–atmosphere connection was an important impetus in the creation of the U.S. National Weather Service in the nineteenth century.

- The pursuit of biometeorology more or less went into remission with the advent of the "germ theory" of disease, but is experiencing a renaissance with concerns about the health effects of climate change.

+ + +

That the weather intensifies or even causes human pain and discomfort, and that it is inextricably tied to our well-being, are truths held to be self-evident through the ages by hundreds of millions of people all over the world.

"We are so certain that weather affects our health that to feel ill or be sick is synonymous with being 'under the weather.'" So stated Frederick Sargent II, a physiologist who was a pioneer in the science that became known as biometeorology.

It has had quite a history.

As a National Research Council (NRC) team stated in its report published at the turn of the millennium, "Health and climate have been linked since antiquity."[1]

Did you know that the atmosphere's perceived effects on the human body were an important impetus for the creation of one of the world's most important scientific institutions, the U.S. National Weather Service?[2]

Biometeorology is a discipline that for a variety of reasons—some explainable, some not—has gone in and out of favor through the centuries, with a variation in intensity of interest to rival the weather itself.

It evidently has entered another up cycle. Biometeorology has experienced a resurgence in the twenty-first century with ever-increasing concerns about how the cosmic consequences of a warming planet and the changing atmosphere might affect disease vectors and assorted viruses, allergies, and chronic conditions.

Granted, those Dr. Algorithm forecasts posted daily by the likes of AccuWeather and weather.com are more about the pursuit of web traffic than concern for public health, but they do speak to how a weather–health connection is so commonly accepted.

That acceptance, said Edwin Grant Dexter, one of the first Americans to examine seriously the weather–body connections,[3] has been a driver of the quest for answers. "The basis of widespread opinion along any line," he wrote at the turn of the twentieth century, "is a fertile field for scientific research, and none more fertile than that of belief in weather influence."[4]

The great nineteenth-century French scientist Claude Bernard, described as the "founder of modern physiology and medicine,"[5] observed, "The conditions necessary to life are found neither in the organism nor in the outer environment, but in both at once."[6]

Certainly, more than one well-credentialed researcher has concluded that the atmosphere's alleged role in various physical torments was the stuff of "old wives' tales," unsupported by hard evidence, and that we have engineered and insulated ourselves from our natural environments. But the credible medical experts and complaining patients who hold that it is the stuff of reality have support from one of history's most famous and oft-cited, yet underappreciated, physicians: Hippocrates.

Not surprisingly, Hippocrates is mentioned prominently in that NRC report.

"Whoever wishes to investigate medicine properly, should proceed thus: in the first place to consider the seasons of the year, and what effects each of them produces," Hippocrates wrote 2,400 years ago in his treatise, *On Airs, Waters and Places*.[7] By then, the concept of the weather–health connection already was at least 2,000 years old, and may well have dated to the dawn of human consciousness.

But it was Hippocrates who extricated the concept from its occult ancestry.

ABOUT HIPPOCRATES

Details of Hippocrates's life are wanting. On that point, historians are in accord. It is known that among his champions were Plato, a con-temporary; Aristotle; and much later, Charles Darwin.[8] Various sources have provided different estimates of his age at death, ranging from 85 to 109. However, it is believed he was born around 2,500 years ago.[9]

Only 60 works credited to him have survived, and it is not at all clear how many of those he actually wrote.[10] The treatises survived a great fire that destroyed the Great Library of Alexandria at the end of the second century CE and were published under the title *Corpus Hippocraticum*. Authorship remains a source of debate, but researchers agree that the Hippocratic influence is unmistakable.[11]

These days, I suspect that many of us confuse Hippocrates with the figures of Greek mythology, like Achilles and Perseus. His legacy is most recognizable in its adjectival form, as in the "Hippocratic Oath." Yet even there, the most oft-quoted phrase from that oath—*First, do no harm*—is actually a misquote. And contrary to popular perception, taking the oath is not a requirement in most modern medical schools.[12]

But, Hippocrates was quite a real person, by all accounts one with a prodigious mind and deeply cultivated sensitivity. Despite some of his misconceptions—reminiscent of some of Aristotle's notions about the atmosphere—he deserves the title of "Father of Medicine," in the view of medical historians. In his book, *Politics*, Aristotle went so far as to knight him "The Great Hippocrates, the wise physician."[13]

He was trained by his "physician–priest" father and his grandfather (perhaps they should be "the grandfather and great-grandfather of medicine"). In that era, the prime qualification for being a doctor was being born into a family of physicians.

Hippocrates was instrumental in changing that tradition: One of his most significant contributions was to establish a medical school where aspiring physicians without family connections could be trained and educated.[14]

He had other profound impacts on the future of medicine.

The standard sculptural portrayals of Hippocrates represent him as a venerable, bearded figure. While they may be essentially accurate, Dr. Fielding Garrison wrote in his seminal *An Introduction to the History of Medicine*, published in 1913, that Hippocrates broke the sculptural mold of the Greeks' understanding of anatomy, which had been limited to "visible or palpable parts." Hippocrates "gave to Greek medicine its scientific spirit and its ethical ideals," said Garrison.[15]

During the apex of Athenian democracy, Hippocrates was alive at the same time as the towering intellectual figures of Western civilization. In addition to Plato, they included Socrates, Sophocles, Euripides, and Herodotus, whom we will meet again in chapter 12. Forgive the pun, but they weren't exactly airheads. (Okay, don't forgive the pun.)

Hippocrates postulated that climate and meteorology were the very starting points for understanding human health. He believed that the body was bonded to the total environment: the air, the water, the topography, even its orientation to the sun. He advised those who chose to pursue medicine to be aware that seasons "differ much from themselves in regard to their changes" and take into consideration "the winds, the hot and the cold, especially such as are common to all countries, and then such as are peculiar to each locality."[16]

Some of the fundamental concepts of biometeorology, although that term wouldn't be used for millennia, have endured.

It is well-established, the NRC authors observed, that "a change in weather can lead to the appearance of epidemic disease. . . . Indeed, the very terms we use today for infectious diseases often preserve ancient notions of disease being caused by environmental factors." They cited "influenza," derived from "influence," and "malaria," from *mal*, or bad, and *aria*, air, not to mention "cold."[17] Of course, the term "cold" describes as many as 200 different viruses[18] and has survived through the ages. (For more on the "cold," see chapter 8.)

DARK AGES

Precisely how weather–body associations might have evolved or devolved during the centuries between the collapse of the Roman Empire and the dawn of the Scientific Age is shrouded in darkness.

But the invention of instruments that could observe and measure natural phenomena—the likes of barometers, which verified the profound insight that air has weight; thermometers; and anemometers and rain gauges—led to quantum advances in the understanding of the behavior of one half of the weather–health connection: the atmosphere. Those advances, in turn, were considered a boon to identifying atmospheric impacts on health. Scientific societies in Italy and Great Britain undertook efforts "to unify and analyze meteorological and medical information," the NRC report noted. "France even required a precise collection of meteorological data by physicians three times each day, on forms provided by the Royal Society of Medicine."[19]

Environmental medicine experienced a heyday in the late eighteenth century. New medical information led to the building of drainage and sewer systems, and the establishment of clear links between certain locations and disease risks. The medical community was actively involved in weather observations in the pursuit of atmosphere–disease connections.

In the New World, meteorology and medicine had a symbiotic relationship. Thomas Jefferson and Noah Webster were diligent in tracking weather and its possible relationship to disease.[20] In 1818, the U.S. Army surgeon general ordered his charges "to record the weather and everything of importance relating to the medical topography of his station, the climate, and diseases prevalent in the vicinity." That practice, which would continue until the 1880s, was aimed at identifying any cause-and-effect relationship between the atmosphere and the soldiers' health.[21]

One of the most important developments in the history of meteorology—the establishment of a U.S. national weather bureau, in 1870—resulted from efforts to make the weather–disease connection.[22]

Ironically, however, that development coincided with a disconnect between medical and meteorological pursuits. Changes were in the winds that would affect the future of biometeorology.

By the mid-nineteenth century, something of a tug-of-war was underway between the weather people and the doctors. The weather folks were envisioning new uses for their observations. With the invention of the telegraph, meteorologists saw opportunities for storm warnings and for using the data to make weather predictions. Physicians remained involved in collecting weather observations, but the data were being conscripted for other uses.[23]

Weather became the subject of a military takeover. The resolution signed by President Ulysses S. Grant in 1870 placed the new national weather agency under the command of the secretary of war. The thinking at the time was that the collection and rapid dispensing of weather data would require military discipline.

Meteorology was marching away from physiology, and it would be decades before the creation of any government health department. The very name of the new weather agency—the Division of Telegrams and Reports for the Benefit of Commerce—suggested it would have nothing to do with physical well-being.[24]

"The military engulfed meteorology," the NRC team wrote. "By 1880 meteorology's early origins in medicine were all but invisible, and medical study of epidemics turned away from prediction toward pursuit of disease agents in laboratory-based investigations." The nature of understanding the origins of diseases was undergoing fundamental change. "The first 'revolution' in epidemiology came after 1882 with Robert Koch's elaboration of the germ theory of disease. . . . Younger, statistically knowledgeable epidemiologists accepted the fundamental determinism of germ theory—that a particular microorganism would be found responsible for any given disease."[25]

But if the Hippocratic ideals went into remission, they did not disappear. For example, in France, between the world wars, "neo-Hippocratics" were embracing a vision of disease that spoke to the relevance of climate to ailments.[26]

Interest in establishing order to the world of biometeorological research gained traction in the 1950s. In the United States, a "Biometeorology" division of the American Meteorological Society was created in 1952. The following year, Dutch geologist Solco W. Tromp traveled to Washington to meet with an American group that included Frederick Sargent. Tromp informed the Americans that he was part of a group of European scientists who gathered in Paris to discuss forming a biometeorology society that would meld biology and meteorology.[27]

The International Society of Biometeorology was formed in 1956, with Sargent serving as its first president.[28]

One indication of the surging interest in the subject is in the publication history of the *International Society of Biometeorology Journal*. Since its beginnings in 1957, nearly 20 percent of all articles that address meteorological variables and their relationship to specific diseases were published just in the years 2017 and 2021, with an uptick evident starting in 2011.[29]

CLIMATE CHANGE, AND COMPLICATIONS

The planet's rising temperatures have been an obvious impetus for the renaissance of interest in the connections between climate, weather, and health and disease.

One of the more obvious potential results of a warming world would pertain to a highly weather and climate–related affliction that affects as many as 2 billion people worldwide—pollen allergies—and the 300 million who suffer from asthma.[30]

Any number of studies published in the last few years have indicated climate change in the lengthening of pollen seasons, and allergists warn that conditions are likely to further deteriorate.

Just how bad might it get? Pollen still is guarding some of its most consequential secrets, and the future may hold quite a few surprises.

SPRING

ONE ACHINGLY BEAUTIFUL SEASON

Among the four seasons in the Northern Hemisphere temperate zone, spring likely would far outpoll the other three in any survey ranking the seasons in terms of qualities that promote well-being.

Personally, I would side with the minority, casting my vote for winter, although it would be a tough choice. I am a fan of extreme times, and spring certainly qualifies, its benign reputation notwithstanding.

Tennyson's famous observation in "Locksley Hall"—"In the Spring a young man's fancy lightly turns to thoughts of love"—has been oft-repeated, and oft-misunderstood and misrepresented.

First, it is indeed a succinct poetical encapsulation of one aspect of what is called "spring fever," a phenomenon experienced by humans and other mammals, the insects, the avian species. I'm astonished at the energy that the birds spend in spring, breeding, nesting, feeding the babies and teaching the chicks the facts of existence.

Spring fever, while not a formal medical diagnosis, is very definitely real.

A seasonal-disorder specialist at Columbia University offered the best quasi-medical definition I've come across, describing spring fever as "a rapid and yet unpredictable fluctuating mood and energy state that contrasts with the relative low" of winter.[1]

It's all about the light, which returns in abundance in spring, says Michael Smolensky, a University of Texas chronobiologist who long has had an intense interest in biometeorology and the co-author of *The Body Clock Guide to Better Health*. And that springtime mood bump is all about the rhythms of the seasons. "We take them for granted," he said in an interview posted on the WebMD site. "People accept the fact that our bodies are organized in space—that our toes are at the end of our feet, and the hairs on our head stand up. But we give little thought to the fact that our bodies are structured in time."[2]

Unfortunately, spring isn't all swaying daffodils and songbirds, as Tennyson's poetical speaker in "Locksley Hall" was well aware. He was rather bitterly recounting a painfully lost love. From there, he transitions to some dark and fascinating ruminations about the state of nineteenth-century life. The poem eloquently captures a long-identified psychological paradox: In spring a young man's or woman's fancies might readily turn to melancholy, for reasons that transcend their understanding.

Among the cherry blossoms, crocuses, hyacinths, tulips, and putting greens, it would be natural to feel strangely energized and inappropriately giddy. How could anyone feel depressed romping through such a bouquet of nature?

A whole lot of people have found the ways.

It might stem from puzzlement at the incongruity over sensing an inner darkness against the background of the increasing light and the reawakening of the environment. That would be a variant of what some folks experience during the holidays, when expectations and childhood memories might demand a response radically at odds with an inner discomfort stimulated by forced jollity.

Spring is a time when circadian rhythms can be disrupted, which sleep specialists warn is a tremendously underrated health issue. Like clockwork, in the leadup to the clocks moving forward an hour in March, the debate over imposing the shortest weekend of the year on much of the world will begin anew.

While bills in Congress and in states across the country have called for year-round later sunsets, the American Medical Association and a phalanx of health organizations insist that daylight savings time is a circadian nightmare with serious health consequences.[3] Based on history, don't expect any changes in the clock-change business.

And in terms of weather, spring in temperate climates isn't all sunshine and harmlessly floating clouds. It hosts the most glorious days of the year, all the more appreciated in contrast to the season it deposes, but winter rarely goes quietly. A cloud covering the sun in April can shave off six weeks from the meteorological calendar. While the weather is trending toward the benign, it also is the peak season for the most violent storms on earth, tornadoes.

The weather's volatility can be a source of physical afflictions. Some complain that it literally gives them headaches, as in migraines.

It is a season of discomfort for those who suffer bone and joint pain, and chapter 2 includes a discussion of one of the grander experiments in medical science that attempted to identify the atmospheric variables that that are sources of torment for arthritis sufferers. This provides an abject lesson in the difficulties of linking weather and health.

And spring is number one for allergies, with trees and grass pollens on the attack, sometimes in concert, all in an attempt to sow the seeds of future generations. Various researchers have concluded that, with a warming climate and conditions favorable for more vegetation, the pollen season will grow longer and more punishing for the millions of sufferers.

Pollen is the subject of the next chapter, and how we know—and don't know—what's in the air.

CHAPTER 1

LOVE IN THE AIR

- As many as 2 billion people in the world may suffer from seasonal pollen allergies.
- The tree, grass, and ragweed seasons take up half the year in some of the world's more populated places, and seasons have been growing longer with the warming of the planet.
- Perhaps surprisingly, pollen is poorly counted in the United States, and it may take some time before the effects of warming on the severity of the pollen seasons are quantified.

+ + +

Early in the morning of the first full day of spring, Dr. Donald Dvorin climbs atop the roof above his office on a quest for evidence. He had set his rooftop trap 24 hours earlier and is about to discover that he has ensnared an alarmingly large legion of captives.

Dvorin retrieves a microscope slide from his Burkard Spore Trap and remands the captives to the laboratory. Magnified 400 times, under the lens the prisoners reveal themselves—thousands of miniature tormentors, sucked into a dime-thin slit a half-inch wide and helplessly bonded to the sticky surface of the glass, each one all of about 1/10,000th of an inch in diameter.

Projected onto a monitor screen, that nondescript sheen that appears on so many mornings atop car hoods and mailboxes and patio furniture comes to life. The assembled ovular cedar pollens and the nearly circular juniper pollens that appear to be stenciled with snowflake-like hexagons are evocative of a Paul Klee painting. Dvorin sees so many of these micro-gametophytes during his examination that he can't count them all in one sitting. He has other duties. He is an allergist. This is his busy season. He has to attend to patients who may well be suffering from what his microscope has revealed. The pollen numbers this day are astonishing.

These are the reproductive emissaries of cedar and juniper, maple and oak. Yes, this is going to be one of those miserable days for allergy sufferers when their respiratory systems are under attack from an all-but-invisible army.[1]

Some scientists believe that he may well be peering into the future.

WHO REALLY COUNTS

The evidence on this particular morning underscores the importance of Dvorin's work. Based on the available research, this daily pollen-counting exercise should become ever more vital to the nation and the world.

That's why Don Dvorin may well be one of the nation's most under-appreciated detectives. Ironically, at a time when they are needed most, he may be among the last of the traditional pollen hunters.

His practice is in New Jersey, whose moniker, "The Garden State," correctly suggests that it is a pollen haven. His observation station is situated not far from the Pine Barrens, that legendary and incongruous 1.1-million-acre preserve of woodland in the nation's most densely populated state. Over 9 million people live in New Jersey. His is one of only three pollen-counting stations certified by the National Allergy Bureau. Coverage in other parts of the country is even sparser.[2]

AIR RAIDS

Of all the connections between human health and the atmosphere, the pollen allergy seasons are among the most evident, and most poorly understood. While some people may file them under "A" for annoyance, and that might include actual allergy sufferers, medical professionals take them quite seriously, with good reason. Asthma, which will be discussed fully in the Autumn section of this book, and pollen sensitivity often have a symbiotic relationship.[3]

Various studies have concluded that in a world made warmer and moister due to increasing carbon dioxide concentrations, more plants will be producing ever-more pollen in their efforts to sow the seeds of future generations.[4]

In addition, tree populations have been growing in ever-expanding urbanized environments, serving a worthy cause. In part, that's the result of attempts to mitigate heat-wave dangers associated with the "urban heat island" effect that has given cities a decades-long head start on worldwide warming. However, more trees mean more pollen.[5] Plus, the extra warmth in the cities appears to be triggering earlier pollen releases, compared with surrounding areas.[6]

All things being equal, the changes in the climate system would mean ever-more raw material to torture the estimated 2 billion people in the world who are allergic to the pollen emitted by trees, grasses, and the homely and ubiquitous "ragweeds."[7]

I have felt their pain, almost every year, from equinox to equinox. In my area, the torment begins with the junipers, cedars, poplars, and maples in March, with the pollen of 15 or more species joining them as the season peaks in mid- to late April. For a period in May the reproductive frenzy of the trees overlaps with that of the incipient grasses, which take over by June. After a pollen hiatus of several weeks, the ragweeds join the party in August and into September.

I have had attacks of sneezing—a hundred or more an hour—that have stripped me of all cognitive function and induced a debilitating physical fatigue. I recall one seizure in a small New York theater where Christopher Plummer was performing *Barrymore* in a one-person show. My sneezing was so disruptive that my wife rather vociferously asked me to leave. By coincidence, we had reservations for dinner that night at Trattoria Dell'Arte, whose signature decorative features include a gallery of famous Italian noses. It was a splendid meal, based on what I could taste of it, and it did not stop the sneezing.

Fortunately, I have never experienced a medical crisis, but some of my fellow sufferers have been known to end up in emergency wards with respiratory complications.

Pollen-related medical costs exceed $3 billion annually in the United States, alone, according to the Centers for Disease Control.[8] That total is almost certain to grow dramatically as the ranks of the allergic are likely to continue to swell. Anything that involves the nonlinear chaotic behavior of the atmosphere is sure to have elements of unpredictability, but the consensus forecasts are calling for harvests of suffering from the seasonal afflictions commonly, and erroneously, called "hay fever."

The weather has almost everything to do with this. It is the brewery and the distributor; a classic illustration of how human health is inextricably tied to the behavior of the atmosphere.

ABOUT HAY FEVER

Just what is this phenomenon called hay fever, what do we know about it, why is it afflicting so many more people, is it any way predictable, and what can an allergy sufferer do about it?

Hay fever has nothing to do with hay. Those who are hypersensitive to pollen are unlikely to feel anything to challenge the respiratory system in the presence of bales of hay.

English physician John Bostock has been credited with being the first to describe classic hay fever symptoms, in an 1819 paper he read before the London Medical and Chirurgical Society, with the page-turning title, "Case of a Periodical Affection of the Eyes and Chest."[9] His symptoms would be ever-so-familiar to millions of allergy sufferers these days, and he reported that they encored every summer, almost like clockwork.

"A sensation of heat and fulness is felt in eyes," Bostock wrote, "first along the edges of the lids, and especially in the inner angles." What follows is "the most acute itching and smarting." Then "a general fullness is experienced in the head," and "irritation of the nose, producing sneezing, which occurs in fits of extreme violence."[10]

Bostock later noted that the only mention he had encountered of similar symptoms was a drive-by reference in William Heberden's *Commentaries on the History and Cure of Diseases*.[11] Heberden's account consists of all of two sentences in a section on "catarrh"—a general term associated with cold-like symptoms.[12] (Note: The Spanish word for cold, catarro, is derived from the same Latin root.) Heberden likely made his observations at least 20 years earlier, since he died in 1801 at age 91.[13]

Exactly who first used the term "hay fever" remains one of the mysteries of the universe.

Bostock mentioned that, regarding the symptoms he had described, "an idea has very generally prevailed, that it is produced by the effluvium from new hay, and it has hence obtained the popular name of the hay fever."[14]

The term obviously had legs and life. If you ever had the questionable pleasure of watching the 1960s TV show *Green Acres*, or at least listening to the opening theme song, you would have heard Eva Gabor complain, "I get allergic smelling hay." That was one of her unsuccessful arguments against forsaking Manhattan for the farm life. Perhaps costar Eddie Albert, whose will prevailed, understood the disconnect.

The hay connection certainly wasn't the first misconception associated with seasonal allergies, which likely have been around as long as plants have emitted pollen and humans have had noses. It is believed that both Julius Caesar and Augustus Caesar were sufferers.[15] Yet the true source of these seasonal afflictions remained a mystery deeply into the nineteenth century.

Leonardo Botallo, an Italian-trained doctor who later became a royal physician in France, evidently was responsible for a significant detour on the road to pursuing the answer. He promulgated the popular misconception that flowers in bloom trigger hay fever attacks.[16]

In 1565, Botallo described upper-respiratory symptoms set off by roses, which he called "rose catarrh." In reality, of course, the roses were wrongly accused. The symptoms in all likelihood were the result of exposure to airborne pollen. Although the term "rose catarrh" has long gone out of fashion, the misattribution has endured, perpetuated in the popular media when actors suddenly start sneezing—without much in the way of conviction—near otherwise innocent blooming flowers.

So, how was pollen identified as the agent? It took some ingenuity and self-sacrifice.

TAKING ONE FOR SCIENCE

In an Allergy Hall of Fame, Dr. Morrill Wyman, a Massachusetts physician, would deserve a plaque. He proved the linkages between allergy symptoms and pollen the hard way, by inhaling ragweed pollen.[17]

Wyman's 1875 paper published in the *Boston Medical and Surgical Journal* tidily summarized what we now know of the career of the ragweed season. "Autumnal catarrh, or 'hay fever,' as it is properly called, sets in toward the end of August as surely as swallows come in spring," he wrote. "Limited to a single month, the disease then disappears, whether treated or not. . . . What is very remarkable, numerous as the sufferers are, they are limited to certain regions. If the sufferer leaves those catarrh regions, . . . he begins immediately to recover."[18]

Wyman suggested that for relief, the victims should head to the elevated areas of the White Mountain region of New Hampshire—never a bad idea, by the way.[19] I can testify personally from many encounters that I never experienced symptoms there, only the majestic beauty of one of the nation's most spectacular natural wonders.

Arguably a larger plaque would bear the name of Charles H. Blackley. In an exhaustive treatise published in 1878, Blackley recounted his extraordinary, and serendipitous, investigations that ultimately established the pollen connection. Suffering from hay fever symptoms for 20 years, Blackley was frustrated by the "scanty literature" on the subject.[20] One summer day he was traveling on a road when a passing carriage stirred up a cloud of dust that induced a sneezing fit. The next day he returned to the same stretch of road and created his own dust cloud in what turned out to be a successful effort to achieve the same results.

Blackley examined a sample of the particles under a microscope lens, aiming to identify what properties in dust that could have ignited his sneezing explosion. Eureka! What he saw "revealed to me the presence of bodies which I now easily recognize as the pollen grains of the grasses."[21]

Blackley's observation of the weather that preceded his encounter almost perfectly described the atmospheric conditions that lead to high-pollen predictions in today's algorithm-driven commercial forecasts: A mix of sun and rain followed by a dry spell. That would have been an ideal sequence for pollen production and pollen flight.

Blackley went on to conduct various experiments on himself, introducing pollen grains to his eyes and nose to induce allergy symptoms. He would invent a popular allergy test. He observed that by applying pollen to his skin, he could produce those telltale reddish "wheals" that reveal a sensitivity. He became an early pollen counter, developing his own primitive pollen traps by attaching coated microscope slides to kites, collecting pollens at different levels of the atmosphere.

Another of his findings was that pollen could travel vast distances on air currents. Charles Darwin was a fan of Blackley's work, and corresponded with him. Darwin hypothesized that hay fever attacks on ships might have been triggered by airborne pollen.

The body of Blackley's work proved unequivocally that pollen was the agent of the seasonal "catarrh." Blackley also took note of what he viewed as a paradox: people in farm country, where one intuitively would expect pollens to be robust, appeared to exhibit no seasonal allergy symptoms. Some modern research strongly suggests that he was on to something.[22]

BOOM TIMES

The causes of hay fever remain a subject of intense speculation and analysis, but unquestionably the incidence of diagnosed pollen allergy symptoms increased beyond exponentially during the twentieth century. Various studies have identified climate change as a primary agent in the twenty-first century, with a warmer, moister world increasing the overall volumes of plant life and, thus, pollen, and lengthening the seasons. The seasonal duration signal already is evident.

However, the upward trend in cases began well before the accelerating warming of recent decades. Western nations experienced a huge bump in hay fever diagnoses from 1870 to 1950.

An "allergy epidemic" was evident in 1935, when Dr. Oscar Swineford Jr., author of the voluminous treatise *Asthma and Hay Fever*, became a professor of allergy and rheumatology at the University of Virginia. The following year he opened a specialty clinic at the university's medical school.

In fact, hay fever had become so widespread in New York in 1946 that City Council approved a bill to eradicate ragweed. It was called

Operation Ragweed[23] and, by the way, it didn't work, a nine-year study documented. The weed was mightier than City Council. The failures were related to the idiosyncratic behavior of the winds, a sobering lesson for those trying to forecast pollen on a day-to-day basis.[24]

What explains the surge in the legions plagued by the symptoms? One major suspect may well be the "hygiene hypothesis." This holds that advances in hygiene have insulated children from exposure to germs that could toughen immune systems and make them less vulnerable to allergenic attacks.[25]

A sneeze is a symptom of a body's attempt to repulse an invader that to others might be accepted as harmlessly as the scent of a rose.[26] Sneezes are "violent respiratory events."[27] One study concluded that the force exerted on the sternum during a sneeze was similar to that exerted on weightlifters during a "moderate-intensity bench press exercise."[28] In a strictly unscientific sense, a sneeze is your body's way of saying, Get the hay out of here. The "ah-choo" is a natural vocalization; the "ah" on the inhale, and the "choo" emitted on the exhalation as the mouth reflexively stays shut during the muscle contraction.[29]

While no one has established precisely why the ah-choo chorus is gaining so many members, the hygiene hypothesis remains an intriguing possibility. It was first articulated in a brief 1989 paper by British epidemiologist David Strachan. He postulated that the increase in allergy incidence might be related to a decrease in germ exposure. He attributed the reduced germ exposure to smaller family sizes and to children having limited contact with animals at the same time that standards of cleanliness were improving. Strachan reasoned that in the old days immune systems were toughened by more germ exposure than in modern times.[30]

In retrospect, a century earlier Blackley had provided supporting evidence to the Strachan hypothesis. Blackley had observed the lack of pollen allergy symptoms among people in farm settings.

Other speculation on the cause of the allergy bump centers on increases in vaccinations and the use of antibiotics, which theoretically could have long-term weakening effects on immune systems. Another suspect would be vitamin D deficiency: Children spend far more time indoors, shielded from the sunshine vitamin, a possibility discussed in chapter 12.

What about genetic changes? Intuitively, those alone would be inadequate to explain the rapidity of the rise in allergy cases since the late-nineteenth century. In the grand scheme, that isn't a long time.

In a study spanning the years 1870 to 2010, a University of Virginia Health System researcher concluded that "in the case of allergic disease, changes in our environment, diet, water quality and personal behavior . . . have played a dominant role." He surmised that "the best explanation for the appearance and subsequent increase in hay fever at that time is the combination of hygiene and increased pollen secondary to changes in agriculture."[31]

Whatever the causes, the numbers are concerning. Estimates vary, but the Asthma and Allergy Foundation of America estimated in 2023 that more than 80 million people have been diagnosed with severe hay fever in the United States alone.[32]

The allergy wave may well be approaching a crest at the worst possible time.

CLIMATE CHANGE AND ALLERGIES

In just about every way, the state of the atmosphere argues persuasively for more abundant production of pollen, with increases in carbon dioxide concentrations, vegetation, temperatures, and rainfall.

Researchers already are seeing compelling evidence of these factors at work in the tree-pollen seasons in the United States, where winters generally have become milder and damper. In temperate climates, trees historically begin dispersing their reproductive seeds in mid-March, but the timetable has been moving up and the season is ending later. An analysis that looked at the 1990 to 2018 period found "widespread and lengthening of pollen seasons" generally, and an overall increase in pollen volume of more than 20 percent across North America. That study, published in 2021, was assembled by several researchers who used a blend of National Allergy Bureau data and computer modeling, and found the biggest increases in the Midwest and Texas.[33]

Affirming those research findings, Estelle Levetin, a highly regarded aerobiologist and professor emeritus at the University of Tulsa, agreed that the pollen seasons are getting longer. But as an academic with hard experience wrestling with the complications of pollen forecasting, over the years Levetin has become my go-to person for nuance. She added an important caveat: The observed lengthening applied to only "some pollen types in some areas."[34]

Similarly, European studies have documented that "some, but not all, pollen types are increasing in severity, season duration and experiencing an earlier onset," British researchers noted in a paper published in 2022. Using 25 years of data, the team examined the recent careers of birch, oak, and grass pollens and concluded that the worldwide warming trend is unquestionably affecting pollen seasons, but their analyses produced decidedly mixed results. Regarding birch, they found "no significant trends for onset, first high day or duration," but warmer temperatures were boosting pollen production in parts of the United Kingdom. Oak seasons "are starting earlier, due to increasing temperature and sunshine totals . . . but are not becoming more severe." As for the grasses, "pollen seasons are neither becoming more severe nor longer." The researchers attributed that flatlining to "a lack of change in some monthly meteorological variables, or land-use change, such as grassland being replaced by urban areas or woodland."[35]

While "there is an urgent need to make the population aware of the effects of climate change and emerging threats," two allergy specialists wrote in a 2022 paper, "there are sparse data on the direct correlation between climate change, pollen seasons, and allergic sensitization. . . . This will require years of detailed pollen counts and meteorological monitoring, with simultaneous recording of the clinical data."[36]

Fiona Lo, a pollen researcher at the University of Washington, in Seattle, observed simply: "Climate change is complex. How pollen can contribute to allergies is also complex." She added that a critical variable to consider might be the most elusive—the human body. Precisely how allergies develop and how an immune system reacts "are all dependent on the individual," she added. Some react to the pollens of multiple species, others maybe only a few.[37]

For the allergic, the level of suffering would be tied to how much of the offending pollens fill the air at a given time.

Predicting a season's intensity, says Tulsa's Levetin, would be a "complex issue." Regardless of changes in that overall pollen volume, daily, even hourly changes in the atmosphere always are going to be critical to how much pollen becomes airborne and where it winds up.[38]

Take the case of the ever-volatile tree season, the first of the three major seasons. While the antecedent winter can plant clues about what is to come, these aren't always especially useful.

Even if a mild, damp winter kick-starts the trees, "if there is lots of rain during the pollen season," said Levetin, "you get wash-out, and the pollen season seems less intense." On the other hand, if a particularly cold winter pushes back the tree schedule, "it may be a shorter season"—say four weeks rather than the typical six—"but it may then appear more intense."

As did Dvorin,[39] Levetin told me that in the case of trees, a wild card may be in play that can have a tremendous impact on the intensity of a given pollen season. For mysterious reasons, the "effort" that trees devote to reproduction may vary substantially from year to year. In short, sometimes the trees behave as though they are not in the mood, the antecedent conditions notwithstanding. Said Levetin, the literature contains reports "that indicate that some species do have a natural cycle of high pollen years—called mast years—and low-pollen years. However, it would take literally many decades of data to show that these cycles were independent of the meteorological conditions."[40]

WHERE THE POLLEN MEETS THE NOSE

In addition to establishing distinct and specific links to the warming of the world, the mysterious behavior presents yet another impediment to generating accurate, consistently reliable daily forecasts. A further complication is that potential pollen volume during a given season can vary tremendously among individual species from day to day.

Dvorin is among the allergists who pay attention to the day-to-day caprice of pollen and have decided that forecasting isn't ready for prime allergy-season time.

So how do sites such as pollen.com, AccuWeather, and weather.com manage to produce accurate forecasts on a daily basis?

The short answer is, they don't. Those health forecasts do incorporate some of the science, but essentially are the works of Dr. Algorithm.[41]

They do look specific and authoritative, with forecasts to the decimal-point level. However, supporting my own analyses, seven U.S. allergy experts scored forecasts in five cities—Atlanta; Seattle; Omaha; Scottsdale; and Washington, DC—and determined that the outlooks often were in conflict with the truth. Their analysis, published in 2024, found that the accuracies were "low" when matched against National Allergy Bureau data. In Washington, accuracy ranged from 42 to 53 percent; 30 to 42 percent in Atlanta; 11 to 34 percent for Scottsdale; 20 to 25 percent for Seattle; and 7 to 34 percent for Omaha.[42]

Those online forecasts are useful for tracking conditions that would be favorable to low or high counts. Just don't take them literally.

In reality, you probably could do this at home if you have access to a credible weather forecast and a reasonably good handle on the peaks and valleys of the pollen rhythms in your area. If you are a sufferer, chances are excellent that you know those seasonal rhythms intuitively.

As for the weather, warm, breezy days with low dewpoints are best for the trees' reproductive missions. Dewpoint is the temperature at which atmospheric vapor condenses into water droplets and is a measure of absolute moisture in the air. As an indicator for assessing pollen-flight conditions, it is far superior to humidity, which is relative to how much total moisture the air could hold at a given temperature. You can find the dewpoint readings and forecasts by the hour for the next several days on any local National Weather Service site, and they are publicly available elsewhere. Dewpoints in the 40s, 30s, or lower might mean trouble for allergy-sufferers.

Dewpoints in April tend to be far lower than they are in summer, and that has a whole lot to do with the trees, whose leafage hasn't yet matured. Leaves retain moisture, then give it back to the atmosphere with evapotranspiration, thus pumping up the dewpoints in summer. In April, the available leaf surface area for holding that moisture is limited.

Additionally, with the general lack of shade, the ground can warm quickly in the April sunshine, which is why parts of the nation often experience extreme April warm spells.

It's almost as though the trees conspire to create conditions ideal for their reproductive strategies.

THE BEST-LAID FLIGHT PLANS

The trees don't always wait for the April warmup, and evidently even less so with rising temperatures. Pollen has exhibited a certain disdain for calendars, and sometimes logic, as Don Dvorin has discovered and rediscovered.

Take that first full day of spring in 2023, Tuesday, March 21, weeks before the typical tree-season pollen peak. While Dvorin has been capturing and analyzing tree, grass, and ragweed pollen for more than 30 years, he was taken aback by what he found on that morning.

"This is extreme," he said. He had not seen pollen levels like this in "several years." This was particularly puzzling. The sample was collected during a 24-hour period that coincided with one of the chilliest days of the month, and temperatures had dropped below freezing that morning.

Dvorin postulated that the mild winter—one of the warmest on record[43]—and almost total lack of snow cover in his area explained the early bonanza. It also happened that dewpoints during the 24 hours in which the sample was collected were in the teens and 20s. The air was ultra-dry. Given the significant warmup that afternoon and similar wind conditions, he anticipated a spell of pollen riot and another extreme count on the following day. Dewpoints again were in the teens and 20s.

Instead, he got another surprise: The counts were 10 times lower than they were on Tuesday. This was further reassurance that he was wise to eschew the forecasting business. Among allergists across the country, he has considerable company.

THE COUNT

While what Dvorin experienced that Tuesday spoke to why he stays away from prediction, it also served as a clinic on why so few of his peers even bother to count the stuff.

His day started at 8:00 a.m. with a trip to the roof of his office building, where he emptied a trap that looks like a miniature, three-legged satellite. Its most prominent feature is a wind-catching fan blade. It also has a rain shield. After removing the slide from the apparatus, Dvorin applied droplets of a dye called Calberla's stain—a combination of glycerol, ethanol, and distilled water—to make the pollen grains visible.

With the magnified evidence, Dvorin was able to estimate how many grains of pollen had passed through a cubic centimeter of air in the previous 24 hours, which would be reasonably representative of the concentrations within a 20- to 25-mile radius of the observation site.

His mission on March 21 was complicated by the sheer quantities.

The volume was so high that he couldn't count it all in one sitting. He had to break off for his day job, which is treating patients with respiratory symptoms, no doubt aggravated by the very pollen he is counting. He returned to the microscope, then back to the patients. Finally, at 3:45 p.m., he sends out a mass email with the headline, "Tree Pollens Extreme High."[44]

For these efforts, he is paid exactly $0.00, and given that they mean time away from his practice—as in the source of his income—his pollen-counting enterprise operates at a loss.

WHO'S COUNTING?

In his 17 years as a pollen counter, Dr. Timothy Craig has encountered his share of surprises. He recalled finding masses of pollen "as thick as oatmeal" in his front-porch trap in upstate Pennsylvania. When a hurricane blew through, he said, "I actually saw pollen grains from palm trees." Of pollen counting, Craig said that overall, "It's actually pretty enjoyable."

A few years back, however, he decided to stop counting. "It's very labor-intensive and time-intensive," said Craig, a professor of medicine and pediatrics in allergy and immunology at Penn State Health Milton S. Hershey Medical Center.[45]

The time commitment was not trivial. Along with the daily trap setting and analysis, the machines used require preparation and maintenance. Estelle Levetin's operating instructions for the Burkhard trap cover 16 pages.[46]

Said Dvorin, "Very, very few allergists want to get into this because of the time factor."[47]

Fiona Lo said she fully understands why the pollen-observation network is so wanting, given that the counts historically have been provided by the likes of volunteer physicians, such as Craig and Dvorin. Naturally, she said, "their main focus is the patients."[48] How might a patient dealing with respiratory discomfort react if, after a 45-minute wait, a doctor said, "Oh I'm sorry for the delay. I was in the lab counting pollen"?

The net result, said Lo and other pollen researchers, is that United States is seriously lacking in reliable, controlled pollen data. "Pollen observations are sparse, sporadic, not-standardized," and "often hard to come by," Lo said.

"It is disappointing that there isn't a better observation system," Lo added. "There is an informal group of us that would like to set up a better open system. However, we need funding and have so far only been able to get small amounts," sufficient only for "small steps."[49]

While automated counting systems hold promise, they remain works in progress, so for now, ground-truth pollen monitoring is left to the volunteers participating in the National Allergy Bureau network. Not just anyone can do this. Those who sign up need certain qualifications and must pass a pollen-identification examination.[50]

In 2024 the network had fewer than 65 certified stations nationwide, and not all of them were submitting daily reports. Fiona Lo's state, encompassing more than 70,000 square miles with a population of 7.8 million, had only two—both on the west coast.

Taken together, Washington, California, and Oregon—an area covering more than 350,000 square miles, with 15 percent of the nation's population—had a grand total of nine.

The only one serving New England, home to 15 million people, is in southern Connecticut, meaning the closest station to Caribou, Maine, would be 500 miles away. The National Allergy Bureau map showed one station in Massachusetts, but the last time I checked for a pollen report there, the site said, "This station has not collected any allergen data."[51]

What would be an ideal geographic distribution of measuring networks? Research into that question also is wanting, but Yolanda Clewlow, who runs the United Kingdom's pollen forecast service, said that a decent "good correlation" in one study identified a 24-mile radius.[52] Obviously, the National Allergy Bureau network would be a tad shy of that standard.

Jeroen Buters, a toxicologist and allergy researcher at the Technical University of Munich, said he was "frustrated" with the lack of confirmed pollen measuring stations in the United States.[53]

Buters was the lead author of a paper published in 2018 that discussed the state of pollen monitoring around the world. "Many apps deliver pollen forecasts," the authors write, "but it is unclear how the data are obtained." They noted that that while several governments around the world, including the United States, require the tracking of chemical pollutants that in some cases might take years to affect people, they appear to care little about pollen that can immediately impact millions.[54]

Among countries having some semblance of observation networks, including Japan, the world leader, and Italy, France, Spain, and Germany, the United States ranked well behind in stations per capita. Japan has a station for about every million people, quadruple the per-capita density of the National Allergy Bureau network.

Buters and his associates mapped pollen-measuring stations around the world, comparing them with standard pollution measuring locations. Their mapping "makes it clear that, in our opinion, biological particle monitoring is a neglected aspect of air quality monitoring. We trust that the map will address this discrepancy."[55]

They may have depicted the discrepancy; "address" would be another matter.

Obviously, a denser network would have immensely added value for verifying forecasts. Knowing how those forecasts performed would be a critical element for fine-tuning them.

Said University of Washington's Fiona Lo: "Increasing the pollen observations would most definitely improve forecasting."[56]

THE FUTURE OF COUNTING

Estelle Levetin and others believe that the automated counting and forecasting devices ultimately would be the great hopes for the future of pollen prediction. Levetin and a research associate, California State

University's Peter Van de Water, had posted mountain cedar pollen forecasts on the Tulsa pollen site for 17 years, but eventually decided to give it up. In making the announcement, they wrote: "We hope that someday, daily pollen forecasts such as these will be replaced by an accurate automated system. However, that type of pollen forecasting is still in the future."[57]

A major step in that direction is underway across the pond. A consortium of organizations across Europe has been working on establishing an automated network.

German researchers, in a 2021 article, saw promise in the future of automated systems, even if they remain a work in progress. "While there might still be a long way until they lead the way in atmospheric biomonitoring," they wrote, "their progress is fast-pacing."[58]

In the meantime, the outlook for upgrading pollen forecasts in the United States has at least one thing going for it, according to Fiona Lo: "Considering that most of the pollen forecasts in this country are not very good, I think we can greatly improve on the current state of forecasting."[59] Donald Dvorin promises that he will keep counting.

If you rank among the legions of pollen-allergy sufferers, along with consulting an allergist regarding medications and shots, here are a practical few tips from the American College of Allergy, Asthma and Immunology.

- **Avoidance works.** In pollen seasons, keep windows closed, and use clean air filters of decent quality in air-conditioning systems.
- **Change clothes.** Pollen can stick to shirts, pants, shoes, and hats. If you've been outside, slip into something more comfortable—and pollen-free.
- **Shower.** Consider adding a shower to the nighttime routine. Pollen can stick to your body and hair and take up residence on your pillow. Pollen is not something you want to sleep with.

CHAPTER 2

DR. HOLLANDER'S CHAMBER OF PAIN

- Over 350 million people in the world suffer arthritis pain, and more than 100 variants and related conditions have been identified, according to the Arthritis Foundation.

- It is estimated that 75 percent of all arthritis patients insist that their pains have definite connections to weather. Doctors say they can't all be wrong, but for a variety of reasons, study results historically have conflicted.

- One pioneer in arthritic research may have taken the greatest pains to find the connections, in the process affirming the complexities of the connection between the atmosphere and the human body . . . and the chasms between medicine and meteorology.

+ + +

The conditions were exquisitely spring-like. The temperature was 76 degrees Fahrenheit, the air comfortably dry and scrubbed clean, not a trace or rumor of pollutants. Those reaping the harvest of this pristinely fresh environment weren't modern-day Wordsworths luxuriating among the daffodils, however. They were captives, having surrendered their freedom to one Dr. Joseph Hollander for the sake of one of the most extraordinary experiments in the history of medical science.

In the history of arthritis research, Hollander's work, though long consigned to obscurity, constitutes an important chapter. His work produced significant results, some of them incidental and invaluable in ways that Dr. Hollander himself evidently didn't recognize at the time. His experimentations encapsulated the complications in matching atmospheric variables with pain, and perhaps underscored the chasms between the disciplines of medicine and meteorology.

It involved a contraption out of a science fiction novel, evoking the Golden Age of Hollywood horror movies. It was called "Climatron."[1]

In his writings, it was clear that Hollander's mission pivoted on a fundamental premise: It is folly to discount the atmosphere's effects on human well-being, akin to ignoring the importance of diet and exercise in an intelligent health program. As the rate of climate change appears to be accelerating, that premise is very much alive and well today among reputable medical professionals and, certainly, among those living with the sometimes debilitating effects of arthritis.

A WORD ABOUT ARTHRITIS

Arthritis is among the oldest documentable ailments and was an impetus for the proliferation of baths in the Roman Empire,[2] according to Hollander, a revered figure in the history of research into the causes and treatment of arthritis.

It is quite a complicated affliction and comes in multiple varieties, including osteoarthritis, the most common type; rheumatoid arthritis; inflammatory arthritis; and gout.[3]

The planet Earth is home to about 350 million arthritis sufferers, says the Arthritis Foundation, of which Hollander was a cofounder.[4]

His interest in arthritis dated to his residency at Pennsylvania Hospital, "where his contact with young patients with rheumatoid arthritis profoundly influenced his career development."[5] As Hollander noted in *Arthritis and Allied Conditions*, while arthritis kills "very few, . . . There is no other group of diseases which causes so much suffering by so many for so long."

As testimony, arthritis patients have posted compelling and, yes, painful descriptions of their ordeals. Here are a few vignettes that have appeared on the Global Healthy Living Foundation's Creaky Joints website:[6]

"It may sound harsh but no one gets what you're going through, not even the doctors," says a California man. "Most types of arthritis simply cannot be explained to someone who has never been through it."

Adds a New York sufferer: "You know that deep vibrating feeling, that ache deep in your bones, that you get after you jump off something tall and land really hard on concrete? That's what I tell people my rheumatoid arthritis feels like."

Says a 50-year-old Australian: "I tell people that my knees . . . feel as if Arnold Schwarzenegger hit them with a baseball bat every time I take a step. They hurt every second, every minute of every day. That's my life, and that's why I don't go for a quick walk to get coffee with you in the morning."

One patient summarized simply: "Arthritis feels like failure."

To at least some extent, Hollander could relate to those comments. He himself was an arthritis sufferer.[7] An age-old complaint among arthritis patients is that weather volatility exacerbates their symptoms, something else that Hollander said he had experienced firsthand.

That volatility and pain are very much year-round phenomena, but spring and fall merit special attention. They are the prime atmospheric battleground seasons in the North Temperate Zone, when polar and tropical air have their most intense skirmishes, all in an effort to balance the planet's temperature. The invasive and lingering effects of winter and summer drive the tempestuous air-mass contrasts so characteristic of these transitional seasons. For most inhabitants of the midlatitudes, spring is the most anticipated season. For arthritis sufferers, spring could well be the cruelest. It evidently has the edge over fall.

In a comprehensive study of 1,665 arthritis patients over a four-and-a-half-year period, Japanese researchers evaluated 10 different measures of pain and found an "increase in disease activity during spring" for every single criterion.[8]

ABOUT OLD WIVES AND THEIR TALES

In describing his reasoning for the grand experiment, Hollander succinctly captured the disconnect between experience and hard evidence: "Most arthritic patients claim that they feel worse before weather changes. This sensitivity to climate variations has defied scientific corroboration or explanation."[9] Note that he used the word "most," which implies a number significantly greater than a mere majority.

The road to his conclusions more resembled that of a Medieval town than a modern turnpike, but even after his first round of experiments, Hollander was confident that he had succeeded in elevating the connection beyond what he termed the "old wives' tale" level. "It now seems reasonable to conclude that weather effect on arthritis is a definite phenomenon."[10]

My mother was one who would not have needed any convincing.

She certainly would have qualified as a master of the "old wives' tale" genre. She was a housewife, who could read the winds: "That air's all in the front. That's rain." That was her way of crushing any hopes for snow, and she often was correct. We lived along the Delaware River in southeastern Pennsylvania, less than 15 miles from Hollander's Climatron. We didn't experience many snowstorms, but on occasion one would come along to liberate us from school. That wasn't likely to happen when the winds were pressing against our front windows, which faced due east, which meant they were importing warmer air from the Atlantic Ocean, 60 miles away, where the water never freezes. Conversely, if the wind was from "Larkin School," to the north and two blocks closer to the Arctic Circle than was our house, we had a shot at snow.

She also was a part-time human barometer who held that her pain was a sure sign of impending atmospheric mayhem. Given her profound suspicions and fears of the medical profession, I cannot imagine what she would have thought of Hollander or his incredible project.

ABOUT THAT "WEATHER EFFECT"

In his experiments, Hollander focused on humidity and air pressure to get at the "weather effect," and with good reason.

Falling air pressure in combination with rising humidity often signal that a storm is coming, and as Hollander wrote, so many arthritic patients associate weather changes with their pains. The humidity speaks to the moisture content of the air, relative to how much water it could hold at a given temperature. Dew point, which will be explored further in the Summer section of this book, is the temperature at which water vapor condenses and comes out of hiding—think of the water droplets that form on a glass of iced tea in summer as the air comes in contact with the cool glass. It is a far better measure of atmospheric moisture, meteorologists agree. It is an "absolute" rather than a "relative" number, but the term humidity is the clear choice of the nonmeteorological world and appears frequently in the medical literature.[11]

Pressure is a more esoteric concept, and arguably the more important agent in the weather–pain relationship. It is a sine qua non in weather and forecasting. It literally is the weight of the air that is pressing upon our bodies.

The higher the elevation, the less dense the air becomes as the number of molecules decreases and, thus, so does the air pressure. While the atmosphere extends upward hundreds of miles to the borders of outer space, about 50 percent of the air molecules are compressed in the lower 18,000 feet.[12]

It is far lighter on a mountaintop than in a valley: The sheer mass pressing on a summit would be dramatically less than that pressing on us at lower elevations. Think of how the weight of water increases as one dives deeper into the sea and how "thin" the air becomes at the highest mountain summits. At the peak of Mount Everest, for example, the air is three times lighter than it is at sea level.[13]

BAROMETER FALLING

When James Joyce mentioned "barometer" in one of his *Dubliners* short stories—"The barometer of his emotional nature was set for a spell of riot"[14]—the image would have been well understood. The stories were published in 1914, a more agrarian time, and deep into the twentieth century the concept would have been at least marginally familiar to the general public. The reference to Joyce's story reappeared 30 years later, twice, in Charles Jackson's *The Lost Weekend*.[15]

In the heydays of Francis Davis and his peers, barometric readings still were getting equal time with temperature, humidity, and wind accounts in media weather reports, which were sure to include the pressure's four-digit readings and rising or falling tendencies. Today, it is the barometer's popularity that is falling, a forsaken arcane measure of the unseen.

Pressure is at once an overwhelming and profoundly subtle force.

The very idea that air has weight eluded the great minds of scientific history, including Galileo's. The insight is credited to his countryman, physicist and mathematician Evangelista Torricelli. In a letter to his mathematician friend, Cardinal Michelangelo Ricci, Torricelli made the cosmic comment: "We live submerged at the bottom of an ocean of the element air, which by unquestioned experiments is known to have weight."[16] It wouldn't be possible to overstate the importance of that insight to atmospheric science—or, evidently, to some arthritis sufferers.

Torricelli invented an instrument that actually could weigh the air, the mercury barometer. In essence, this was a scale in which a column of mercury in a glass tube would rise or fall depending on the weight of the atmosphere immediately overhead. He opted for mercury over water because mercury is 13.6 times denser; using water would have required a column 13.6 times taller, or higher than a three-story building. Smart choice.

These days, a variety of barometers that don't rely on the highly toxic mercury are in use, but the readings still are commonly expressed in inches of mercury in the imperial system. On average, at sea level—the boundary between us and the briny deep—the pressure makes the mercury rise 29.92 inches. Relaying the pressure at sea level would be a whole lot tidier than attempting to describe it at different elevations.

So why not just tell us what the air weighs? For one thing, it's doubtful that it would be meaningful to the public.

The mass of air surrounding the planet is more or less constant, but its weight shifts in ever-restless waves from place to place. That 29.92-inch figure equates to 14.7 pounds of air per square inch pressing upon us if we're roaming around near sea level. That amount varies within a virtually imperceptible range, from approximately 14.5 to 14.9 pounds. Imagine if on a sunny morning, a TV or radio weather person announced: "The air this morning weighs 14.8 pounds per square inch at sea level, and it's getting heavier." Aside from the cryptic element, such a pronouncement might even confuse the viewers and listeners since it would go against popular perception, depending on the conditions outside.

While people are prone to say the air is "heavy" when it's gloomy and raining, the opposite is true. The air likely would be heaviest in the heart of a fair-weather Arctic high-pressure system whose descending currents would be repelling clouds and precipitation. Conversely, it would be lightest in the ascending currents near the centers of strong storms: Precipitation falls when warm air is forced upward and changes state by condensing in the cooler higher atmosphere.

How might pressure variations affect an arthritic person? Again, you will find any number of conflicting studies out there. As a medically reviewed summation on the WebMD website explained, "It could be that when the cartilage that cushions the bones inside a joint is worn away, nerves in the exposed bones might pick up on changes in pressure." Or perhaps barometric changes cause tendons to expand and contract.[17]

No one is quite sure. "We still don't know whether it is one particular feature of the weather or a combination of features that matters," wrote Robert H. Shmerling, senior faculty editor at Harvard Health Publishing.[18]

Whatever the physical explanations might be, generations of arthritis sufferers long have insisted that the pain associated with weather changes is in their bones—not in their heads.

SUPERSTORMS

In his attempts to quantify this perceived relationship, Joseph Hollander, the man behind the Climatron, created something akin to arthritic superstorms.

But first, before he varied the controls, he had to get his subjects accustomed to "normal" life in his chamber, which they occupied two at a time for two to four weeks. The lodging was akin to a spacious submarine and, fittingly, had a submarine-like hatch for exiting and entering. It was equipped with beds, easy chairs, and a bathroom; more or less a comfortable studio apartment, with a 270-square-foot combination bedroom/living room.

The chamber was housed in a reinforced concrete structure built inside what was the Rehabilitation Building of the University of Pennsylvania Hospital by one of the nation's premiere climate-control companies, Charles S. Leopold, Inc. The firm had designed air-conditioning systems for the U.S. Capitol, Madison Square Garden, and the New York Stock Exchange, among other buildings.[19] The chamber had quite a complex air-circulation system, including a "double turbo-blower" that was an exhaust system and a supplier of fresh air. Air was passed through an "absolute filter" to remove particulate matter, and then a charcoal filter for deodorizing.[20]

For the first five to seven days that the subjects occupied the chamber, Hollander kept conditions constant. The room temperature was set at 76 degrees Fahrenheit; the relative humidity, 30 percent; and the pressure, 30.00 inches of mercury, close to the typical sea-level barometric reading. All in all, the conditions paralleled that of an ideal May afternoon in Philadelphia. While they acclimated themselves, the patients learned how to measure and record the observations that would be critical to the experiment.

Four times a day they were to take, and then note, their body temperatures with an oral thermometer, along with the timing, location, and intensity of any joint stiffness or pains. Twice daily, they took blood pressure and pulse readings, and were required to measure and weigh accurately their intake and output of fluids.

When the adjustment period ended, the pressure was on. Without the knowledge of the subjects, Hollander raised the pressure to 31.50 inches over a two-hour period. He then lowered it to 28.50 inches during the next four hours.

After "a day or two," Hollander varied the humidity levels in a range from 30 percent to 80 percent in 12-hour cycles—variations not all that unusual in the real atmosphere over Philadelphia—while holding the pressure steady. "Within the next day or two," he increased the humidity and lowered the pressure in tandem.[21] The aim was to replicate the conditions typically preceding the arrival of storms when the air would be swelling with moisture and becoming lighter. On 40 occasions the patients were subjected to the humidity/pressure change combo.

And how did they feel as the manmade storms were approaching? Among the eight with rheumatoid arthritis, within a few hours of the onset of the simultaneous changes, seven experienced "significantly worsening" pain in 29 of those 40 episodes. The symptoms disappeared rapidly when conditions were returned to normal states.[22]

His findings spoke to his hypothesis that the *changes* wrought the discomfort—not the absolute values of the humidity or pressure.

PUZZLING IMMUNITY

Why the seven didn't experience that "worsening pain" in 11 of the episodes was a mystery. And along with those mixed responses, one mystifying result—and this was the type of cosmic hiccup that has bedeviled the pursuit of connections between the weather and human physiology—was the fact that one of the subjects was absolutely asymptomatic.[23]

That was all the more perplexing given the pressure extremes. While the changes in the humidity levels in such short periods weren't all that unusual in the real atmosphere over Philadelphia, the air pressure variations were almost unimaginable, which may well speak to the fact that medicine and meteorology are quite different disciplines.

Dr. Hollander was a brilliant medical researcher, but he clearly was not a weatherman.

To those unfamiliar with barometric readings in meteorology, a change from 31.50 to 28.50 inches of mercury—a 10 percent difference, or a drop in the weight of air from 14.9 pounds per square inch to 14.5— in a six-hour period might seem like the height (or depth) of subtlety. Such a swing in the atmosphere in the outside world, however, would be the raw material of weather history.

BOMB ATTACK

Meteorologists have a fitting term for a storm that blows up rapidly—"bomb." To qualify, the pressure in the vortex of a storm has to fall

24 millibars, or about 0.7 or more inches, in 24 hours. That would be 0.03 inches an hour.[24]

John Gyakum, the McGill University atmospheric scientist who specializes in extreme weather and is credited as the cocreator of the term, told me that indeed such a dramatic change would be hard to imagine, at least on this planet.[25]

That change from 31.50 to 28.50 would essentially be a nuclear meteorological bomb, a deepening rate of 0.33 inches an hour, 10 times the bomb threshold. That's something out of *The Day after Tomorrow*, a movie in which the climate system changes in more or less two hours. Never has the official pressure reading at the official Philadelphia measuring station exceeded 31 inches.

The climate chamber wasn't the atmosphere that we all inhabit. In the chaotic real atmosphere, all the variables "are changing simultaneously," points out University of Manchester's David M. Schultz, lead author of a smartphone study involving over 10,500 UK participants who suffered chronic pain, arthritis in more than 70 percent of the cases. The study did find some correlation between lower barometric pressure and pain. However, it also noted that in a review of 43 studies, 29 of them found some correlation, but some them said the results were "not clinically important."[26]

Harvard's Robert Shmerling is among those who has given the possible weather–health connections serious consideration. In a 2019 blog post, he wrote: "When my patients tell me they can predict the weather by how their joints feel, I believe them." That said, until he encountered "more compelling" evidence, "I remain a skeptic about the weather/arthritis connection."[27]

But in a 2020 post, he appeared to have become less of an agnostic. "Having reviewed the studies, I find myself not knowing how to answer my patients who ask me why their symptoms reliably worsen when the weather is damp or rain is coming, or when some other weather event happens. I usually tell them that, first, I believe there *is* a connection between weather and joint symptoms, and second, researchers have been unable to figure out just what matters most about the weather and arthritis symptoms or why there should be a connection."[28]

In 2024, Shmerling said in an email message that while he doesn't "scour the literature regularly," as an arthritis specialist the possible weather–arthritis connections remain of high interest to him. And since that 2020 post, he said, "I haven't seen anything new that's compelling or that changes my view."[29]

In the conflicting findings, what is evident is what Schultz called a "lack of consensus" on how individual variables affect pain.[30]

And in the case of Climatron, given the extreme changes of the variables in such a compressed period, the uneven responses spoke to the fact that, while Hollander could control the weather inside that climate chamber, he couldn't control that other critical variable—the human body.

Dr. William C. Shiel, a cofounder of MedicineNet, now part of WebMD, and other arthritis specialists do agree that many patients experience a worsening of symptoms with changes in the weather.

So much for the easy part. "In one room I may have a patient complaining that last week, just before it rained, her joints began aching and now that the weather is warm and clear she feels better," he said in an interview posted on MedicineNet. "Simultaneously, in the next room, a patient tells me that her joints are far worse today after it rained last week! What do I do with this information?"[31]

COMPLICATIONS

The contradictory results perhaps aren't surprising, given that the Arthritis Foundation holds that more than 100 types of arthritis and related conditions have been identified. In addition to the most common variants, they include "metabolic arthritis," caused by excess uric acid in the body. The acid can form "needle-like crystals" in joints, explains Shiel, setting off episodes of stabbing pain. While "it appears that there is some evidence that the symptoms of certain people with arthritis are influenced by changes in the weather," said Shiel, "this is not true for all people with arthritis, nor is it predictable what type of weather alterations will bother people."[32]

Moving to a more benign climate likely would be no panacea: Shiel pointed out that Arizona has a generous population of arthritis specialists. Various studies have concluded that the transplanted body develops a new relationship with the local climate, in effect redefining what constitutes a change in the weather.[33]

In the universe of arthritis, no one size ever would fit all. This was painfully evident in the results of Dr. Hollander's grand Climatron experiment.

After this first group of experiments, Hollander noted that in a solid majority of the environmental-change exposures, the patients had experienced symptoms. Hollander persevered in his quantification efforts, cycling patients through the Climatron for several years. The triumphs and exasperations expressed in his writings perhaps encapsulate the promise and pitfalls of biometeorological research. In the end, Hollander concluded, the pursuit had been worth the endeavor. "Despite the innumerable frustrations and hours of wasted effort to achieve data reportable only in a 'Journal of Negative Results,'" he wrote in 1968, "the 'old wife's tale' that arthritics can predict the weather has been statistically proved."[34]

Hollander presented his early findings to the American Rheumatism Association at Chicago's legendary Edgewater Park Hotel in 1962.[35] In the opinion of at least one of his subjects, Hollander had gone through an immense amount of trouble to affirm the obvious. "Hell, why did you have to go and spend all that money," he is reported to have said. "I could have told you that years ago."[36]

Here are some expert tips from the Arthritis Foundation.

- **Medicate.** Take the prescription or over-the-counter drugs suggested by your doctor to help control inflammation and pain.
- **Scale down.** Excess weight puts more pressure on weight-bearing joints and increases pain. Plus, fat tissue can send out chemical signals that increase inflammation. Avoid processed foods and sugary drinks.
- **Keep moving.** Along with helping to control weight, walking, water aerobics, pedaling a stationary bike, and other activities can reduce joint pain and improve flexibility. And you might even sleep better.

- **Stay positive.** A positive attitude can be invaluable to fighting pain. Don't focus on the pain. Do things you enjoy and spend time with family and friends to put pain in its place.[37]

CHAPTER 3

HEADACHE, THE "ALARM SYSTEM"

- Headaches share one trait with common colds: Just about everyone gets them. More than 200 types of headaches have been identified.
- Perhaps surprisingly, a headache in some ways can be a "good thing," and might even be an ancestral gift.
- Research has produced convincing arguments and counterarguments, but unquestionably some headaches have everything to do with weather and seasonality.

+ + +

Among all the nouns in the English language, few can rival "headache" for negative associations. An odious chore at work, an irritating colleague, a stack of bills, a leak in the roof, a shot transmission, any other annoyance you care to add to the litany of human burdens and discomfort would qualify for acceptance into the phylum "headache."

When was the last time you heard any human being say something remotely positive about a headache?

In my case, the last time occurred during a conversation with Dr. Paul G. Mathew, an assistant professor of neurology at Harvard Medical School, and a national expert on the subject. Paul Mathew has one perspective that differs quite radically from the popular perception.

"Headache," he says, can be "a good thing."[1]

At least a majority of the time, and for most of us. We might even view it as an ancestral gift. A headache might be a clue that some insidious infection or potentially dangerous ailment is brewing in our bodies, or it might be the body's way of saying: Back off, skip last call. Get some sleep. Go move your body. Stop thinking about work. Shelve the anxiety, and perhaps heed the wisdom often wrongfully attributed to Mark Twain: "Most of the things we worry about never happen."[2]

A headache can serve a purpose, says Mathew, who holds fellowships with the National Board of Physicians and Surgeons, the American Autonomic Society, and the American Academy of Neurology. "Headache is like a car alarm system that warns people something is wrong, like an infection or a tumor. Fortunately, for the vast majority of headaches, there is no sinister underlying pathological cause, but rather the alarm is going off inappropriately."[3]

For some, however, in particular migraine patients, the system can go haywire. "Now imagine if in addition to the alarm, the radio turns on, the air condition turns on, and the car starts going in reverse. This is what happens with a migraine, and the person may experience light sensitivity, sound sensitivity, and nausea/vomiting in addition to headache pain."[4]

Unfortunately, for the tens of millions who live with headaches, including migraine and the "cluster" variant, the pain transcends what most people experience.

Why is it that the worst of the headaches target only certain people, and just what is known about this "alarm system" we call a headache?

SIMPLY COMPLEX

The headache shares characteristics with not only the common cold, but also asthma. As asthma, it is set off by certain "triggers." Like the cold, it is amazingly common, affecting perhaps 96 percent of the population.[5]

When I told Dr. Mathew I was part of the privileged 4 percent, he was incredulous. He insisted that in some point in my life I must have experienced at least a dull aching sensation that in the court of common sense would qualify as a headache. Under his cross-examination, I acknowledged that I wasn't so special after all, that on occasion I did belong with the 96 percent, and that what I had experienced would fall under the technical definition of headache.

Stripped to its essentials, "A headache is pain or discomfort in the head, scalp or neck," according to the National Institutes of Health.[6] However, that simple definition belies layers upon layers of complexity. At last count, in 2018 more than 200 variants of headaches[7] had been identified by the International Headache Society's Headache Classification System.[8] Yes, that's enough to give you headache.

Not surprisingly, given such a myriad of types, headaches have quite a buffet of possible causes, or what the medical professionals call "triggers."

While the causes certainly are "multifactorial," Mathew[9] and other experts are certain that weather is a major factor, even if the evidence is wanting and/or conflicting in the scientific literature.

WHEN CHANGE ISN'T GOOD

Atmospheric volatility is a big player, said Mathew, and that spring and autumnal air-mas moodiness is no friend of the headache sufferer. The atmosphere tends not to have much of an attention span in spring and autumn. A tranquil day that is breathtakingly gorgeous and balmy can assume a wholly different personality when one of those Canadian-born fronts plows through a region and imposes a whole new regime of conditions.

Why might that affect headache patients?

Cold and warm fronts form and move along the boundaries of clashing air masses. Cold fronts typically travel from west–northwest to the east, and their warm counterparts, from the south–southwest to the north. On conventional weather maps, the two-dimensional illustrations of these ongoing air-mass battles classically show an extensive cold frontal boundary and its foreboding dragon-like tail trailing vertically from a powerful cyclone. To its right would be the more or less horizontal, bowed warm front, with benign looking semi-circles. It may appear that the cold front is moving it along as though it is pushing a cart. It is, in most cases. In these frequent air-mass skirmishes, the smart money is on the cold front.

What the weather maps don't capture is that these confrontations are three-dimensional melees, extending miles into the atmosphere. Warmer air rises over the colder air mass, and a cold front launches any warm air in its path violently skyward. The results often include violent storms, the lightning flashes and the cracking of thunder almost evoking an atmospheric version of a headache.

The approaches and passages of fronts are characterized by dramatic changes in barometric pressure, or the weight of the air. Ahead of a cold front, that warm air tends to be less dense and lighter; behind the front, it is denser and heavier.[10]

As we noted earlier, the particulate matter tossed airborne and then thrown to the ground can be hazardous to those with respiratory ailments. The air-mass changes also can aggravate arthritic pain, and that very much appears to be the case with headaches, at least for some of the victims.

For example, Mathew said that in the turbulent fall of 2023, the effects were evident with "rapid shifts in barometric pressure and precipitation, . . . pouring rain one minute, bright and sunny the next minute. It was wreaking havoc on close to 70 percent of my patients."[11] Perhaps more than coincidentally, in line with Mathew's estimate, a National Headache Foundation survey found that nearly three-quarters of sufferers are convinced that weather is behind their pain.[12]

Migraine sufferers, or migraineurs, in particular "commonly describe" weather as a trigger.[13]

ABOUT MIGRAINES

Migraines belong in a special subgroup of headaches, although those who experience them might apply a more pejorative adjective.

Up to 40 million Americans experience migraine, with women outnumbering men by about three to one, according to a StatPearls review published on the National Library of Medicine website and updated in August 2023.[14]

Indian researchers have described migraine as a "genetically influenced, complex disorder and a neurological illness" that may affect over a billion people worldwide and is a "leading cause" of disability.[15]

Even the word itself has a complex etymology. It is derived from Greek, Latin, and French words that more or less mean pain on one side of the skull,[16] and that doesn't begin to tell half the story.

"When you hear 'migraine,'" says the American Migraine Foundation, "you may think of a severe headache. But migraine is more than just a headache: It is a debilitating neurological disease that comes in many forms, both with and without headache." The foundation lists multiple subtypes of migraines. They include:

- **Chronic.** Up to 5 percent of the U.S. population experience headaches in as many as 15 days a month in three different months annually. If migraines occur on 8 or more of those 15 days, the condition is considered chronic. Pain intensity can vary tremendously, and it is possible that patients may mistake the lesser variants for tension or sinus headaches.

- **"Let down."** Stress is a well-known migraine trigger, as it is for so many other conditions, but for migraineurs, it isn't always over when it's over. Sometimes, when a stressful situation passes and a person is about to relax, another wave of pain develops. That may be related to changes in cortisol levels. Stress stimulates cortisol production, and relaxation has the opposite effect. It's possible that the fluctuations are behind the secondary wave of pain.

- **Aura.** About 25 percent of migraine sufferers experience unpleasant visual sensations, such as seeing dots or sparks, or feel a tingling or numbness on a side of the face, usually signaling that an attack is coming. Most patients avoid aura and skip right to severe head pain, often accompanied by light sensitivity and/or noise.

- **Retinal.** In rare cases, patients temporarily lose vision in one eye. This is most likely to happen among women in their twenties and thirties. The migraine affects vision only in one eye, as opposed to the typical aura phase, which affects both eyes.[17]

Just why so many people have to endure migraines has long been a subject of research and speculation. But one credible hypothesis may surprise—yet make sense—to some migraineurs.

Once upon a time, or times, the migraine may have been a life-saver.

ANCIENT WEATHER CONNECTIONS

The first available records mentioning migraines date back 4,000 years, to the civilizations of Mesopotamia.[18]

But humans almost certainly experienced migraines long before they started recording their sensations for posterity. It is entirely possible that in prehistoric times, human wiring was ultrasensitive to signals that the atmosphere was concocting something life-threatening.

Credible research points to the possibility that migraine may have had its origins in evolutionary advantage. According to Mathew, "When humans were hunter-gatherers, and a horrific storm was approaching, developing a migraine may have caused them to seek refuge in their cave even before there were any visible signs of the storm."[19]

Notably, all the genes that cause a migraine make the sensory organs and cortex of the migraine sufferer hypersensitive. "In a state of hypersensitivity, the brain could recognize external threats easily," opined Korean neurologist Dong-Gyun Han in 2020. Therefore, it makes intuitive sense that these "evolutionary advantages" would be a source of migraines. "Migraine was designed by natural selection to solve problems faced by our ancestors in the distant past."[20]

MIGRAINES AND FORECASTS

Granted, this type of personal sensory prognostication would have been in vogue a few eons before we had all those annoying smartphone emergency alerts to tell us about impending weather hazards anywhere near our zip code.

But today, some patients are convinced that their migraines are a source of weather-predictive powers, said Dr. Mathew. They will say, "It's going to rain later day. . . . I feel a migraine with certain features coming on."[21] Sometimes their inner forecast sensors are as reliable as weather-prediction models, while on other occasions, they can prove as fallible as those computer models. Weather fluctuations have been known to come and go with no symptoms experienced by migraine patients.

Why the inconsistency? "The problem is that people frequently look at triggers in isolation," explained Mathew. Other things may be happening in their lives besides experiencing the approach of a cold front that might make them well prepared to weather, or avoid, a storm. "When there is a weather event on the horizon, they may have had adequate sleep, be under no significant stress, and be well hydrated, satiated, so the combination of triggers is relatively low."[22] Conversely, in the case of a migraine that came on with the storm, it is possible the patient had slept poorly due to a coughing child or had been under significant stress from a project at work.

The complexity of migraine pathophysiology and variability from person to person, makes it especially challenging in terms of identifying and disaggregating triggers, said Mathew. This identification process can be daunting for both the patient and the physician.

Unfortunately, that can result in some patients' misidentifying the trigger. "There are people who say, 'I had a slice of pizza, and then I had migraine, so I am never eating pizza again.'" While it is possible that something in the pizza may have played a role, "pulling an all-nighter and having too many beers before the pizza probably played bigger roles."[23]

We don't need a migraine to know which way the wind blows these days, but the evidence strongly argues for weather–headache connections.

What variables might be in play?

Barometric Pressure

As with arthritis, one prime suspect in the onset of migraines is barometric pressure, the invisible force that is so poorly understood by the public, and perhaps is one of the clues that alarmed our friendly hunter-gatherer ancestors.

In chapter 2, we discussed how a pioneering specialist went to extraordinary lengths to establish links between barometric pressure, which these days is an under-the-public-radar phenomenon, and arthritic pain. He had some successes and more than a measure of frustration.

The establishment of any links has been even more elusive in the case of migraines. But if all the suspect weather variables were forced to participate in a police lineup, chances are that a lot of the migraineurs would identify barometric pressure as the guilty party.

The weight of the air, which the barometer measures, shifts in subtle ranges around the average—14.7 pounds per square inch—with the more radical changes occurring with approaching and departing weather systems.

When the air is dry and cold, it tends to be heavier. In the rain-producing rising air of storms, it is lighter, thus, the barometric readings are lower. As a storm approaches, the barometer falls.

Said Dr. Cynthia Armand of New York's Montefiore Medical Center, our sinuses are really air pockets "at equilibrium" with air pressure. That balance is disturbed when the pressure changes, and, thus, that can precipitate a migraine.[24]

In a Japanese study of headache patients during the approach and passage of a typhoon in 2014, 25 of 34 of those with migraines reported pain with decreases in air pressure. (By comparison, of the 28 in the study with tension headaches, only six reported symptoms.)[25]

Results in other studies have been far less robust regarding the role of pressure. As Canadian neurologist Werner J. Becker observed, "Patients have implicated all of high barometric pressure, low barometric pressure, falling barometric pressure, rising barometric pressure, and any change in barometric pressure."[26]

Atmospheric Moisture

A number of studies have tied migraines to "humidity," a somewhat misleading term since humidity is relative to how much moisture the atmosphere can hold at a given temperature. The higher the temperature, the greater the moisture capacity. It would be more precise to reference the absolute moisture content of the air, but most of the researchers aren't meteorologists, so we'll let them get away with "humidity."

A broad Japanese study published in 2023 that involved more than 15,000 migraine patients concluded that along with pressure, high levels of "humidity" and rainfall correlated "with an increased number of headache occurrences."[27]

A more focused analysis by a group of U.S. researchers involving patients in Berlin found that a significant subgroup of migraine sufferers is "highly sensitive" to weather elements that include atmospheric moisture and temperature.[28]

Just how a moister atmosphere could contribute to the onset of migraine is yet another enigma. It is worth noting, however, that heat and humidity—and they have histories of overlapping into fall, even in colder climates—may be conducive to dehydration, a migraine trigger.[29]

Temperature

In a survey of 66 migraine patients in Taiwan who kept diaries for a year, more than half reported that temperature changes affected their symptoms.[30]

In Germany, an analysis involving 4,700 migraine reports found a "significant increase" in symptoms with radical temperature changes.[31]

A team of genetic researchers uncovered possible evidence of a link between the cold and migraines. They found a genetic variation associated with migraines that is nearly 18 times more common in Finlanders than in people of Nigerian descent. The researchers suggested that thousands of year ago, this variation gave people in the far northlands an advantage in adapting to cold. Unfortunately, it also made them susceptible to severe headaches.[32]

Light

On a picturesque January morning, even the average winter-phobe can savor the spectacular beauty of the sun beaming on a snow cover. For most, the scene induces a sense of tranquility. For migraine sufferers, it can induce pain.

"For many migraine patients, natural light is the enemy," says the American Migraine Foundation, which ranks light, along with weather conditions, among the top 10 triggers. Migraines are "closely linked" with "photophobia," an extreme sensitivity to light, says the foundation.[33]

WHICH ENVIRONMENTAL FACTOR MAY BE THE MORE DOMINANT IN TRIGGERING MIGRAINES?

Barometric pressure? Humidity? Temperature? Or light? Research and patient reports have made cases for each, and the evidence makes a reasonable argument for all four, separately or in combination, contributing to headaches in at least some patients.

And so do other factors, identified and unidentified. The complexity of migraine pathophysiology and variability from person to person presents an ongoing challenge to identifying causes, said Mathew.[34]

"DRILLING THE INSIDE OF MY BRAIN"

Whatever underlying factors may be precipitating their migraine episodes, the descriptions of the torment that patients endure are almost too painful to read.

Wrote one patient, in a post that appeared on a site maintained by Harvard Medical School, "On the good days, the pain was just a mild throbbing sensation. Other times, there was a general sense of an ever-tightening pressure. . . . On the days when I couldn't get out of bed, it felt like someone was tightening screws into the sides of my head and pounding a hammer above my left eye."[35]

Of her migraine with aura experience, a woman on the Migraine Trust website described the frightening "unpredictability" of the attacks: "One moment I can be cycling through a field, the next, I can't see." She has experienced "a variety of neurological symptoms," such as losing feeling in parts of her body. Then comes the pain. "It is not just a headache. It feels like somebody is drilling the inside of my brain."[36]

Added an Australian sufferer, "My bad migraine pain is like continuously hitting your thumbnail with a hammer, but on the side of your head at the temple."[37]

A doctor at the University of Utah School of Medicine, who one day would end up treating people with migraines, recalled that his symptoms started when he was about 14 years old. "I was hiking with my dad, and I started to go blind. Then I had this weird pain and nausea. I thought I was going to die!"[38]

CLUSTERS, AND CLOCKWORK

But the migraine is not the king of head pains, says Mathew. That distinction may rightfully belong to the cluster headache.

Although migraine can significantly impact function, cluster headaches tends to be more severe, and can cause greater disability.

The headaches typically occur during the change of seasons, with attacks lasting 15 minutes to three hours, and occurring up to eight or more times per day around the same time of day during an active cluster cycle. For example, a patient can report having attacks at 3:00 a.m., 5:00 a.m., 9:00 a.m., noon, 5:00 p.m., and 8:00 p.m. on a daily basis during an active cluster cycle, which can last several weeks or months.

Mathew said cluster headaches affect about 1 in 1,000 people, with men outnumbering women by about three to one, the opposite of the male–female migraine ratio. "Cluster headache is believed to be the most painful condition to affect human beings," he said. "It's been described as a hot poker stabbed through the eyeball."[39]

An International Headache Society study of more than 750 cluster patients showed a clear correlation between headache episodes and radical temperature changes, with the transition period from fall to winter being especially troublesome.[40]

What is unmistakable, said Mathew, is that cluster headaches do adhere to specific, tyrannical rhythms, with onsets painfully predictable for those who get them. "Cluster headache is incredibly circadian," he said. The headaches "typically occur during the change of seasons, . . . the same time of day during the cycle. . . . That's on the nose, and on the clock."[41]

As with migraines, clusters are associated with photophobia, that ultrasensitivity to light that comes with its own array of effects.[42]

While no one is quite sure why clusters are so tied to the seasons and so predictably recurrent, one plausible hypothesis is that the predictability is related to the changing of the light as the time between sunrise and sunset undergoes changes, disrupting sleep cycles.[43]

SCIENCE AND REALITY

A disconnect between what people experience and what serious research can document is a leitmotif of biometeorology.

One of the overwhelming frustrations in headache research—and this is applicable to arthritis and any number of other conditions that may have weather–body connections—is that studies have shown most everything. And their conclusions often don't mesh with what patients perceive.

Said Dr. Mathew, "What makes it into scientific journals and what occurs in clinical practice—real-life experiences of patients—can be very different from the perspective of physicians like me who have been treating patients for many years."[44]

In an essay published by the International Headache Society, Dr. Becker, the Canadian neurologist, addressed the question: "Weather and migraine: Can so many patients be wrong?"

His short answer: "Personally, I don't think so." Becker itemized several obstacles that have frustrated the pursuit of establishing indisputable relationships between migraine and specific weather conditions. For one, he is with Dr. Mathew on the "multifactorial" aspects. "There are at least 60 trigger factors for migraine," he noted.[45]

In an expansive survey of more than 1,000 migraine patients, researchers ranked weather as number 4 on a list of possible triggers, after stress, hormone issues, and hunger.[46]

So, how, then, to tease out weather from other triggers in a given pain episode?

Echoing Mathew's observation, Becker said that a given weather trigger may hold its fire on the next go-round. "This will make correlation of headache attack onset with the occurrence of any particular trigger difficult," he wrote.

Another variable is the timing of weather changes, which don't always happen abruptly. Fronts and storms travel at different rates.[47] Becker cited the example of chinook winds, those powerful warming gusts that develop on the lee side of mountains.[48] Those winds have been linked to migraines.

Half of the chinook-sensitive patients in an Alberta study reported symptoms with the onset of the winds—and half reported feeling pain after the passage!

"The mechanisms by which weather-related factors or indeed any trigger factor precipitates migraine attacks are not understood," Becker summarized. "Therefore, it is difficult to know which of numerous weather factors to focus on."[49]

But, yes, based on his experience and that of other experts such as Mathews who have dealt with patients in the real world of pain, the atmosphere has been a contributor to headaches, and will continue to be.

OUR LIFE-SUPPORT SYSTEM

So often overlooked is that, in an era when we spend so much time indoors, insulated from the environment, we have a long and interactive relationship with the atmosphere, our life-support system. In the words of climatologists Stephen Schneider and Randi Londer, "In a sense, climate and life grew up together."[50]

The disconnects, the conflicting perceptions, the puzzling findings, the inevitable frustrations in the pursuit of headache causation all make sense when we remember that fundamental truth: the human body and the environment meet at a cosmically complicated intersection.

Becker observed that, yes, migraine patients who indict the weather have good company in arthritis patients. "In arthritis, there is also evidence that the patients are right, but that the response to weather is very individual and likely not present in all patients.

"Where does all this leave us? Perhaps it is still best to listen to our patients."[51]

+ + +

Here is some headache-relief advice from Mayo Clinic specialists:

- **Standard tension headaches.** Over-the-counter remedies usually are effective.

- **Migraines.** Along with medications, rest in a dark, quiet place. Use hot and/or cold compresses. And a small amount of caffeine may provide relief.

- **In general.** Just about all of us experience headaches and most of them aren't worth your anxiety. If you find that they interfere with your personal activities or work life, see a physician.[52]

ADDENDUM
A COSMIC DUST-UP

A serendipitous encounter with road dust led to one of the most significant insights in the history of pollen-allergy research.

Here is the remarkable personal account of Charles Blackley from his book, *Hayfever: Its Causes, Treatment, and Effective Prevention.*[1]

+ + +

I had several times noticed that dust could at certain times of the year produce some of the milder and less-marked symptoms of hay-fever, but there was this peculiarity about these attacks, that generally they came on only during the time that hay-fever prevailed (and then as exacerbations) or immediately after the hay season was over, but rarely, if ever, during winter or early spring.

There was also another peculiarity which these attacks had, namely, that they were more fitful and more ephemeral, coming and going in a more irregular and transitory manner than the ordinary attacks of the disease ever do when they have once set in. At first I was considerably puzzled, and was unable to account for the fitful appearance and departure of the symptoms. I also noticed that the attacks were more frequent whenever I had to pass through any dusty lane in the country, when the hay had been recently all gathered in. I was consequently inclined to think . . . and others have since thought, that common dust was one of the causes of the disorder.

In one of the earlier years of my attacks, before I had made up my mind to follow out a systematic course of observations on the subject, and when I was just getting free from the disease (about the middle of July), I was out in the country and had to walk through a lane which was apparently not often used for the passage of vehicles. A carriage which passed me at a rapid rate raised a cloud of dust in which I was, for a time, completely enveloped and compelled to inhale pretty freely before I could get out of it. A very violent attack of sneezing immediately came on, and continued at intervals for about an hour. As I had to pass over the same road on the following day, I determined to see if the same result would follow by disturbing the dust voluntarily. I found that I could bring on the symptoms in this way to the fullest degree of severity.

The first examination of the dust under the microscope, made with that which had been scraped from the road and examined in its dry state, did not show anything very special. A second examination of the upper layer of dust mixed with glycerin was more successful, and revealed to me the presence of bodies which I now easily recognize as the pollen grains of the grasses.

So far as I can now recollect, the weather during this season had been very favourable for the rapid growth and flowering of grass—first a few hours of rain, then a day of sunshine—and when this got to be nearly ready for cutting, and before the period of flowering was gone by, the weather had settled down so as to give three or four weeks without any rain.

With the help of subsequent experience, it is not difficult to see why such a season as I have described should have given rise to a condition of things which would quite account for the symptoms from which I suffered.

SUMMER

HOT TIMES, AND GETTING HOTTER

As a kid I awaited summer with an inexpressible anticipation. That had a lot to do with my adversarial relationship with school, which I referred to as a minimum-security prison with a liberal weekend-furlough program. Like the rest of the inmates, I viewed the end of the school year as a development that was too good to be true.

It wasn't just that.

It was summer, at last! It was sunshine, promise, baseball instead of teachers' dirty looks. Cold Cokes, Popsicles, ice cream, outings to lakes and pools, culminating with two magical weeks at the shore in August. All this without homework.

I suspect that my attitude wasn't in any way special.

I also suspect that, for the overwhelming majority of my peers, anticipation far exceeded reality. That was certainly the case with me.

I would say my euphoria had a shelf life of about 48 hours, give or take an hour, from the point that the prison gates opened. It was something far deeper than boredom. In my memory, I spent so much of my summers in a state of discontent. Those sensations have never gone away.

SAD-NESS

During a July heat wave, I received a message from Dr. Norman Rosenthal, the psychiatrist who first identified the condition known as seasonal affective disorder. Through the years I've had many conversations with him about SAD, and you will find more about Dr. Rosenthal's work in the Autumn section. The most rapid loss of light of the year occurs in the fall, even though SAD is often called the "winter blues." For about 5 percent of the population, the resulting enervation and psychological stress can lead to clinical depression.

That July Rosenthal was about to have a book published, *Defeating SAD*, which addressed a topic that especially piqued my interest—the summer variant of SAD.

Some patients actually have a negative reaction to longer day lengths that can interfere with sleep and, of course, to heat and humidity, explained Kelly Rohan, a SAD researcher at the University of Vermont. I certainly can relate to both, growing up in a house without air conditioning and inhabiting a body made all the hotter by extreme sunburns. Rohan says summer SAD sufferers tend to see everything with dark-colored glasses. Instead of; say, savoring the liberation from school, they may be overwhelmed by a sudden lack of structure.[1]

Yes, the last thing the world needs is another diagnosis, and this one is ripe for skepticism. But Rosenthal said the summer version of SAD actually presents a greater suicide risk than the winter variant, and various studies have documented that suicides become more common in the warmer times of the year.

Tonya Ladipo, a Philadelphia therapist who works primarily with Black patients, observed that sometimes it's important for a person to say, "It doesn't bother you, but it bothers me."[2]

The summer isn't all "lazy, hazy crazy days of soda and pretzels and beer." In so many ways summer is a season of paradox. It is a time when the atmosphere turns the outdoors into a grand open-air theater, while brewing more than a season's worth of hazards.

It is the season when we have the most opportunities to absorb natural vitamin D, but accepting that invitation indiscreetly can result in skin damage with long-term consequences.[3]

Based on the counsel of my own dermatologist, Dr. Seana Covello, who has done her best to save my sun-sensitive skin, I concluded that in my youth I did almost everything wrong. She advised me not only to use 30+ SPF sunscreen, but to cover up as much of my body as possible. That's whether resting or in motion. "It's the same sun," she said. "It's the same ultraviolet radiation." She had no advice for keeping seagulls from attacking our food.[4]

Picnicking outdoors is one of the great pleasures of summer, and, unfortunately, so are incidents of food poisoning. Very simply, food spoils more rapidly when the temperatures rise.[5] Summer also is the prime season for hiking in the woods and on mountain trails and for gardening, and also the prime time for Lyme disease-bearing ticks seeking blood meals.[6] It is a season for dining on patios and admiring the fireflies electrifying the nightscape, and for mosquitos and their vector-borne diseases, the likes of dengue fever and West Nile virus—whose pathways may be changing with the climate.[7]

Summers seemingly have been displaying a certain disdain for astronomy in parts of the nation, with hot weather starting before the solstice and lingering past the equinox. That's a boon to warm-weather activities, but the atmosphere appears to be becoming steamier as the world warms. Researchers have been warning that the warming will extract health consequences.

For all the hazards attendant to summer, the one that turns out to be the most challenging and the deadliest is the most obvious—heat. It can present tremendous physical challenges, especially to those unaccustomed to it.

CHAPTER 4

HOW DOES IT FEEL?

- With more than 2.5 million sweat glands at the ready, the human body has remarkable powers to adapt to heat and sultriness, but it does have limits.
- Decades-long efforts to capture how the atmosphere makes us feel under a given set of weather conditions and to quantify the risks remain works in progress.
- Along with heat and the moisture content of the air, ground-level ozone and other pollutants are serious threats to health in the summer, and wildfire smoke has become a growing hazard, even in the Eastern United States.

+ + +

While attending a soccer match among the usually temperate hills of central Pennsylvania in early June 2023, on what became a historic summer, Paul Pastelok had a disconcerting sensation. Something in the clinging, steamy air was draining the life from him, as though his consciousness had sprung a leak. To be clear, Pastelok wasn't playing. He was watching the match as his son and other sweaty teenagers romped around the field. When he got home, Pastelok, an otherwise healthy middle-aged man, realized he had left something vital at the stadium: His energy. He recalled thinking, "I can't do anything." The next day, he declared to his wife, "I'm not going back out there."[1]

I wondered just how bad it was "back out there" in State College, Pennsylvania, that afternoon. Yet when I scanned the National Weather Service's hourly reports, I had to triple-check the numbers. During that game, the "heat index," or what the temperature theoretically feels like, never got above 82. If Pastelok were any ordinary resident of the atmosphere, one might conclude—to invoke a pejorative weather-related cliché—that perhaps he had been out in the sun too long.

But Pastelok is a senior expert meteorologist at AccuWeather Inc., one of the world's largest private weather services, and is the company's— and one of the nation's—leading long-range forecasters.

I would have written off his experience as hallucinatory, except that I'm certain, based on firsthand information, that his experience was quite real. I had a similar sensation when I went outside that day, about 150 miles away from that soccer match.

I thought I was coming down with the flu—and likely so did millions of other people in the Mid-Atlantic and Northeastern United States. Yet nowhere in those dense population centers did the heat indexes remotely approach the triple-digit figures associated with potentially dangerous levels of discomfort that develop when the atmosphere becomes swollen with those invisible broths of water vapor (a phenomenon better described by the "dew point," rather than the misleading "humidity," as discussed below).

Those days were weeks away, as was the peak of the lung-challenging ground-ozone season, not to mention the arrival of vast quantities of acrid Canadian wildfire smoke.

As I set out to explore what in the known universe was going on with my body and those of my fellow sufferers, an anecdote I never forgot came to mind. Ironically, it involved something that happened in winter.

"THE PROFESSOR WAS RIGHT"

In the Northeast, winter was reawakening, but the weather on the afternoon of December 29, 1979, was splendid on Florida's west coast.

"It was a beautiful day in Tampa."[2] So pronounced Richard "Batman" Wood, a linebacker for the Tampa Bay Buccaneers, a team that had experienced a remarkable turnaround from the era of "Throw McKay into the Bay." That was a popular saying among fans of a hapless team that had lost 26 consecutive games—14 in their inaugural 1975 season, and 12 more in 1976. The "McKay" in question was Hall of Famer John McKay, who had been one of the most successful college coaches in history before his ill-fated debut in the professional ranks.

The Bucs stuck with McKay, and in 1979, just three years after earning a plaque in the hall of ignominy, McKay led them to a 10–6 record, a division title, and the playoffs. The Bucs' first-round opponent was the Philadelphia Eagles, a team with an elite defense, an all-pro running back, and an explosive passing game. Philadelphia, viewed as a legitimate Super Bowl contender, finished with a better record than the Bucs, but since the Eagles didn't win their division, they had to travel to Tampa Bay for the first-round matchup. Still, even on the road, the Eagles were solid favorites.

For the Eagles, the result was disastrous. Led by Doug Williams, a Black quarterback—a rarity in those days—on its opening drive Tampa Bay bullied its way 80 yards downfield and into the end zone. The Bucs built a 17–0 lead and advanced to the next round. The Eagles took a plane back to Philadelphia and their crushed, stunned, and bewildered fans.

One person who saw it all coming was an unlikely source. He was a professor and dean of science at Philadelphia's Drexel University, a premier science and engineering institution. His name was Francis Davis, who also was among the nation's first television meteorologists. The diminutive Davis, all of 5-foot-5 with hair parted sharply in the middle, wasn't exactly the prototypical beer-drinking, barebacked, green-bellied Eagles fan as seen on TV, but he sensed something that Eagles coach Dick Vermeil and the journalists who covered the team had missed.

Davis had contacted Vermeil—who, by the way, almost always was accessible and responded to phone messages—and implored him to take the team to Tampa Bay the week before the game.

Vermeil decided not to take the advice; after all, that would have meant being out of town for Christmas for the players and the entire staff.

He acknowledged later that he had made a mistake. "The professor was right," Vermeil told me in an interview.[3]

Why was Vermeil wrong?

The weather that day in Tampa Bay was about as good as it gets on December 29 anywhere in the country. The temperature at game time was 74 degrees Fahrenheit, and the atmospheric moisture content appeared to be at a comfortable level.[4]

But to Vermeil's surprise, during the pre-game warmups his players, frighteningly, appeared gassed. He feared that some of them actually might pass out. After the game started, they were lifeless through most of the first half, and were never able to catch up to their aggressive opponents. By contrast, Bucs players bolted out of the gate and were able to exploit the Eagles' listlessness on their way to what was a major upset.

Davis had pressed the point to Vermeil that the Bucs would have a tremendous ally that day—the atmosphere. As pleasant as the weather was for the fans and everyone else in the Tampa Bay area, it presented a formidable challenge to the bodies of the Eagles players, who were traveling from the onset of winter in the Northeast. The Bucs, by contrast, were right at home and had not been on the road for three weeks, when they had lost to the San Francisco 49ers.

Davis had warned Vermeil that he was walking into an atmospheric ambush that could be averted by having the team spend several pre-game days in Tampa Bay to become acclimated to the radical change in atmospheric conditions.

It is impossible to know precisely what role these conditions played in the outcome, but the Eagles evidently were affected by a malaise similar to that experienced by Pastelok and millions of other residents of the Eastern United States on that June day in 2023.

The human body has remarkable powers of thermal adjustment, but its limits are severely tested when it is subjected to rapid changes in the composition of the atmosphere in which they live and breathe. This is especially problematic for people who live in midlatitudes and experience abrupt changes in air masses, but it can be an issue even for Floridians and Southern Californians.

Temperature, alone, rarely is the cause of heat stress.[5] All thermal comfort—and discomfort—is relative. That's something the National

Weather Service now recognizes and, on an experimental basis, is incorporating into its heat-warning system.[6]

ODE TO THE HYPOTHALAMUS

We associate summer with rest, relaxation, vacation, leisure—a season in which it seems that our bodies are reading the cues of the languid air and opting for well-deserved breaks. Yet, while the heat and the vaporous broth that surrounds us in the not-so-great outdoors on a hot summer afternoon may create the sensation that our bodies are savoring their torpor, something quite remarkable is happening inside us.

Some medical professionals have described the process in eloquent detail; however, most of us remain oblivious to the internal miracle that unfolds when we confront the atmosphere's thermal challenges. I would liken our bodies to fine restaurants, where you can enjoy great meals and never see all the sweat and hard work that go into preparing them in the kitchens. The chef in this analogy would be the amazing hypothalamus, a portion of the brain no bigger than an almond. It connects the brain to the spinal cord, and relays messages to the rest of the body to regulate vital life functions.[7]

Internal thermal controls are critical, and just as the atmosphere tirelessly attempts to keep the planet near a constant temperature, the human body endlessly works to balance heat gains and losses to maintain our body temperatures within about a degree of the 98.6 Fahrenheit ideal.

Take a minute to thank the mighty hypothalamus, located roughly in the middle of the brain. Its workings have been likened to that of a home thermostat. The hypothalamus monitors a body's internal temperature to ascertain how it compares with the normal value. If the temperature is lower than it should be, the hypothalamus will order the body to produce some heat, and to chill when it's hotter than it should be.[8]

It is quite an operation. Bodily thermoreceptors alert the hypothalamus that it's getting too toasty for comfort. The hypothalamus will respond by ratcheting up the blood flow to the skin, which stimulates glands that produce a liquid composed of water, salt, and potassium that we inelegantly call "sweat." On average, people have more than 2.5 million of these glands. During heavy exercise, a body can lose more than a half-gallon of water, and better than a quart.[9] (No wonder they keep telling us to hydrate.)

The glands are all over our bodies. You likely have encountered the term—or even experienced—"sweaty palms." In fact, for some people this is a serious disorder, called palmar hyperhidrosis,[10] which can be socially embarrassing. What is it about the palms that makes them so prone to sweat, especially in times of emotional stress? It so happens that they are the venues for some of the highest concentrations of sweat glands in our bodies. Other prime areas include the soles of our feet.[11] You might say that on sultry days, our sweat glands wait on us hand and foot to power an uncanny natural air-conditioning system.

Water is a magical substance, and sweat cools us by disappearing. When water evaporates—changing state from liquid to vapor—it has a cooling effect on the skin. The overheated body supplies heat energy to transform sweat to vapor.

This evaporative cooling effect is evident when one steps out of a hot shower, and the vanishing water cools the body dramatically and instantly—sometimes too much for comfort. We reach for the towel for warmth as much as for dryness.

Dogs should be jealous: Sweat is way more efficient for cooling than panting. That doesn't mean it is effortless, though. Cooling the human body when those heat indexes climb into triple figures requires a tremendous amount of unseen internal labor.[12]

That process is all the more labor-intensive when the temperature and moisture content of the atmosphere spike precipitously. That was the unpleasant experience that meteorologist Paul Pastelok, this author, and likely tens of millions of Easterners had on that first Saturday of summer in 2023.

In early July of that year, when it turned seriously warmer in the Mid-Atlantic and Northeastern states, the National Weather Service took the extraordinary step of advising Easterners that it was perfectly normal to feel listless as the weather turned abruptly sultry, after a month dominated by some of the coolest June weather in decades. "We've heard from a lot of people that it seems exceptionally uncomfortable," the agency said in a social media post. "While the heat and humidity we have presently is nothing unusual, June was unseasonably cool. . . . This kept many from adapting to the higher heat level of July."[13] Understandable. With a modest amount of adaptation, our natural air-conditioning system typically functions fine.

But what happens if that system breaks down?

NO SWEAT

That summer staple, "it's not the heat, it's the humidity," speaks to an imprecision that meteorologists long have accepted with a degree of equanimity. More properly, it is the heat and the amount of water vapor in the air that determines the level of discomfort.

The "relative humidity" is a misleading quantity for the very reason that it is, indeed, relative. It indicates how saturated the atmosphere is at a given time relative to how much water vapor it could possibly hold.

It is common for the humidity to spike to near 100 percent on a summer morning when temperatures are around 70 degrees Fahrenheit. On an afternoon when the temperature approaches 100 degrees Fahrenheit, however, rarely would the humidity exceed 50 percent. The air can hold far more moisture at 100 degrees, compared with 70.[14]

Think of a 12-ounce glass containing 6 ounces of water. It would be holding 50 percent of the water it could, relative to its capacity. By contrast, a 6-ounce glass with 6 ounces of water would be 100 percent full.

The Better Measure

"And pray, who are you?"
Said the violet blue
To the Bee, with surprise
At his wonderful size,
In her eye-glass of dew.[15]

In describing the moisture in the atmosphere, meteorologists far prefer dew point. It is an absolute measure, the temperature at which invisible water vapor condenses into droplets and comes out of hiding, as it did upon the violet in the John Bannister Tabb poem referenced above.

If the dew point is 70 degrees Fahrenheit, and if you take a glass of iced tea outside, the outside of the glass will bleed water because the air around the rim has been quickly cooled to the dew point.

Nevertheless, the National Oceanic and Atmospheric Administration's (NOAA) heat-index calculators are pegged to temperature and humidity.

To know what you're in for when going outside on a hot summer afternoon, you can save a few seconds by checking the dew point. Even assuming you're reasonably healthy, if that dew point is around 70 degrees or higher, your body's air-conditioning system is in for a battle. Dew points above that level are indications that the atmosphere is brewing a water-vapor consommé. A saturated atmosphere inhibits sweating, which means sweat drips off or, worse, sticks to your skin. It is no longer a precious substance, but what we might indelicately describe as an unpleasantly insulating slime.

This is a source of discomfort for some, but for the elderly and those with background health conditions, it can be dangerous, even life-threatening. The cooling system breakdown can cause heat cramps or heat stroke, "basically organ failure as the body begins to cook itself."[16]

UNREAL NUMBERS

A doctor I once interviewed, who specialized in the effects of weather on human physiology, remarked that the government's heat index works only for people who are outside in the shade, and naked. He was half-joking, but his comment spoke to the fact that the question posed at the start of this chapter—How does it feel?—is perhaps a surprisingly complicated one to answer.

Nor is quantifying the dangers imposed by various degrees of heat a simple exercise. The human body indeed has elements of elegant chaos that can match the atmosphere's.

Human bodies come in all shapes, sizes, ages, and conditions. What single index could capture the effects of heat on a well-fed construction worker who is outside jackhammering a sidewalk, a wispy jogger, a toddler, and an elderly person recovering from surgery?

SHADY ESTIMATE

The most popular metric that attempts to do all that is the National Weather Service heat index. It is a calculation based on the temperature and the levels of atmospheric moisture, either in terms of the dew point or that more popular measure, relative humidity.

Moisture is a major driver of the index. For example, around 10:00 a.m. local time on a July day during the endlessly searing Southwest summer of 2023, it was 103 degrees Fahrenheit in Phoenix. In Miami, at 10:00 a.m. local time on the same day, it was 90 degrees. The heat index in Miami was 103. The heat index in Phoenix was . . . 103. The dew point in Phoenix was 53. In Miami, it was 77.[17]

While the temperature and moisture level are the only two variables in the heat index equations, the computations are rather complicated. If you want to try your hand at computing a heat index, here is a formula, with "T" representing the temperature in Fahrenheit, and R, the humidity:

$$\text{Heat Index} = -42.379 + 2.04901523T + 10.14333127R - 0.22475541TR - 6.83783 \times 10^{-3}T^2 - 5.481717 \times 10^{-2}R^2 + 1.22874 \times 10^{-3}T^2R + 8.5282 \times 10^{-4}TR^2 - 1.99 \times 10^{-6}T^2R^2$$

And that's actually a dumbed-downed version.

The heat index is based on the pioneering work of Robert G. Steadman, an Australian who was a textile researcher at Texas Tech University. He published a two-part treatise on the subject. His computations involved 15 parameters—including a person's size and clothing, and the wind speed—all with assumed values for the purposes of simplification. The "person" in question would have been 5-foot-7 and weigh 147 pounds, wearing a short-sleeved shirt and long pants and walking 3.1 miles per hour at a moderate pace in the shade, in a 6-mph wind. Through the years, researchers have refined his work in their efforts to fine-tune their estimates of degrees of comfort.[18]

The figures vary from location to location, but a common trigger for a National Weather Service "heat warning" is an index of 105 degrees Fahrenheit. That's an attention-getting number, and sounding the alarm bells constitutes the real value of the index, said weather service warning specialist Sarah Johnson.[19]

In a place where 105 degrees would be the threshold, 104 degrees would not quite make the cut and would qualify only as "advisory" level heat. Realistically, it would take quite a special person, even one who happened to be 5-foot-7, weighed 147 pounds, and was walking 3.1 mph in the shade, to perceive the difference.

Either way, 105 or 104, the probabilities are excellent that the subject would quickly develop a rather unwelcome case of sweaty shirt.

A BETTER WAY?

Coinciding with increases in global temperatures and warmer summers, researchers have been looking at alternatives to the heat index.

A personal favorite is the so-called wet bulb globe temperature, or WBGT. That is not a household acronym and has survived well beneath the public radar, but this index has a long history and has been used for years by the military, the Occupational Safety and Health Administration (OSHA), organizers of running and other sporting events, and college athletic programs.

Unlike the heat index, the WBGT aims to measure the impacts that heat will have in direct sunlight. It considers wind, dew point, cloud cover, latitude, and the level of solar radiation, which is ever-changing as the planet executes its 584-million-mile orbit around the sun.

Like so many innovations in the history of meteorology, the WBGT has a military connection. It was first used in the 1950s as a tool to mitigate heat stress among new recruits in the U.S. Army and Marine Corps, whose "boot camps" were located in the Carolinas. The Army and the Marines had based guidelines on temperature and humidity, but the compliance standards were costing them in lost training time. That changed in 1956 when they switched to the more-comprehensive WBGT.[20] For the military, a color-coded "flag scale" ranges from green to black, with green recommending "discretion" and black calling for all physical training to be "suspended." OSHA has similar guidelines.

The American College of Sports Medicine also has adopted the WBGT, with separate charts for "continuous activities," such as distance races, and "intermittent activities," such as football and soccer. The continuous category has four levels of recommendations, and the intermittent has six, with the most severe recommending cancelling the event.

One problem with the WBGT is that the general public might be unimpressed by the numerical values. The cautionary advice on the American College of Sports Medicine scales begins with a WBGT in the 60 degrees Fahrenheit range, and in the 80s on the OSHA and military scales. The most dangerous levels on all the charts are 90 degrees or greater.[21] Those numbers aren't nearly as attention-getting as, say, a heat index of 110. For precision, the WBGT is superior, but not for messaging.

The heat index isn't going away, although the weather service has been experimenting with a different warning system called HeatRisk, which speaks to what Pastelok and millions of others experienced that first weekend in June 2023, and the defeated Philadelphia Eagles in December 1979. It ranks heat hazards on a simple five-tier, color-coded scale from green, or 0—as in, "come on out"—to magenta, or 4—best to stay inside with the air-conditioning running.[22]

HeatRisk has advantages over both the heat index and the WBGT in that it considers whether the heat is unusual for the time of year, how long the heat is going to last, nighttime temperatures, and Centers for Disease Control data on heat-related impacts.

Missing from the calculations are humidity levels and dew points, the hourly data for which have been lacking in the West and other countries, the weather service says. Instead, the index uses a proxy measure to infer those values, looking at unusually high overnight low temperatures and tight spreads between daily lows and highs. Both are indications that the air is well-stocked with moisture. Water vapor inhibits daytime heating from escaping into space, making nights warmer. During the day, water vapor tends to put a cap on temperature since the sun has to devote some of its energy to evaporating water rather than heating. So, an abnormally small difference between a daily high and low temperatures is an indicator of high dew points.

As of the summer of 2024, the HeatRisk index remained in that "experimental" development stage and was not being used in populated areas in the sometimes-sweltering East. Nicholas Carr, a meteorologist who is a Utah veteran of the weather service, told me, "Some people think it's over-aggressive at times." Eventually, the weather service might use it nationwide.

Will HeatRisk be able to determine how you and I actually feel when we are outside at any given moment on any given day? It would take more than an index to answer that question, and to pinpoint just how dangerous the heat can be to each of us individually. The various indexes, however, can provide us with guidelines.

Intuitively, at least to me, it would make sense for the government to develop an all-hazard index that would account for both heat and air quality. Heat warnings are issued separately from foul-air alerts: Meteorology and the control of environmental pollutants are very different sciences, with specialties within specialties. Don't bother asking a hurricane expert to talk about winter storms or tornadoes.

Interestingly, Sarah Johnson, the National Weather Service warning specialist, pointed out to me that the HeatRisk color-coded system is a page from the air-quality ranking system.[23] I was encouraged to see an analysis by U.S. and Mexican researchers, published in 2022, that attempted to look at how such a combined index health-hazard index might work, and delighted to see that they agreed it would make sense. As the study authors wrote, "Simultaneously considering air pollution and heat is . . . underscored by the fact that high air pollutants and extreme heat often co-occur, in part due to the role of photochemistry in ozone formation."[24]

An analysis of deaths in several French cities during the epochal European heat wave of 2003—which will be explored in detail in chapter 6—concluded that ground-level ozone did make some contributions to the staggering death tolls.[25]

It would be surprising if that were not the case. Ozone, which in the old days was better known as "smog," is something your lungs can happily live without. The American Lung Association calls it "one of the least well-controlled pollutants in the United States and also one of the most dangerous."[26] When ozone is inhaled it can damage lung tissue, and with frequent exposure, the effects can worsen. The National Aeronautics and Space Administration (NASA) has noted that ozone "is such a highly reactive gas" that it is used to disinfect drinking water and has been implicated in 250,000 deaths annually worldwide.[27] In its spare time, by entering the tiny leaves of plants, ozone can decimate billions of dollars' worth of crop yields.

THE OZONE BAKERY

Summer heat provides a vital ingredient in the ozone recipe. Ozone is a byproduct of fossil fuel emissions, but isn't emitted into the air directly. Rather, it is the result of the chemical reactions among the sun's ultraviolet radiation, nitrogen oxides, and volatile organic compounds.[28]

The U.S. Environmental Protection Agency (EPA) reports that the nation has made some progress in reducing ozone levels in this century. Overall, ozone concentrations trended downward about 20 percent from 2000 to 2022. They reached a nadir in 2020, the peak of the COVID-19–related lockdowns, and ticked upward in 2021. However, the downward trend resumed in 2022.[29]

Another NASA study found that ozone reductions resulting from a decade of emissions controls may have saved hundreds of lives in the Eastern United States during the brutally hot July of 2011.[30]

In recent years, particularly in 2023, people in the Midwest and East have experienced another pollutant, one with which the American West is more than well-acquainted.

SMOKE GETS IN YOUR SKIES

While wildfires are an essential component of ecosystems, a warming world and drought conditions have conspired with a variety of other factors, including land-use changes, to increase wildfire incidence, coverage, and intensity in the Western United States.

The smoke is more than an aesthetic blight. It has been tied to any number of undesirable health outcomes, including the aggravation of respiratory conditions such as asthma, and cardiovascular diseases. The long-term effects of continued smoke exposure have been a subject of research.

Easterners have learned in recent summers—particularly in the summer of 2023—that wildfires are more than a Western problem.

"When you live on the East Coast and hear about the West Coast wildfires you assume those wildfires won't have any personal effect on you other than empathy for those in the paths of the fires. Think again." This was posted in 2017 on a NASA web page that included a satellite image of Western smoke riding the upper-air, west-to-east jet stream winds across the country.[31]

Four summers later, smoke from the Pacific Northwest, borne on upper-air winds, congested skies across the contiguous United States all the way to the Eastern Seaboard. In the East, said Tom Kines, senior meteorologist with AccuWeather Inc., "to have it this thick, it doesn't happen very often."[32]

It was in June 2023, however, that the smoke hazards hit home to Easterners. In June, smoke affected air quality from Minnesota–Dakotas borders to the Florida Panhandle.[33] An unusual—perhaps unprecedented—outbreak originating in eastern Canada smothered the Northeastern U.S. population centers with acrid smoke in the lower levels of the atmosphere. The odor was so powerful that it set off fire-related 911 calls.[34]

The Philadelphia Phillies experienced the first "smoke out" postponement in the team's 140-year history. Remarked Phillies first baseman Rhys Hoskins, "It's pretty wild. . . . It's terrible." New York City experienced its worst air-quality level on record. It was a summer in which Canadian wildfires burned 10 times the normal acreage, according to government data.

Said Kines, "It's like all of Canada is on fire."

HEAT. OZONE. SMOKE. THE FUTURE OF SUMMER?

Precisely how much climate change had to do with stoking the Canadian smokehouse is uncertain, but unquestionably it was a factor, said John Gyakum, an atmospheric scientist at Montreal's McGill University whose specialty is extreme weather. Quantifying its effects is another matter. "Whenever you have natural disasters you have more than one factor," he added.[35] The smoke invasion of the Northeast resulted from an idiosyncratic alignment of circumstances. The acreage burned in Quebec was 500 times the normal amount. Why did so much in eastern Canada burn?

In smoky New York City, Amy Freeze, a Fox Weather meteorologist, said the fires in Quebec may have had a contribution from an unlikely source—a beetle infestation. The blight resulted in more dead, combustible wood that was all the more burnable due to a bone-dry May. A tropical storm remnant the previous September took out trees in Nova Scotia that may have supplied more kindling. Another element was what Gyakum called "a peculiar structure to the wind." Ordinarily June winds blow from the west and southwest; however, persistent breezes from the north channeled the smoke southward.[36]

It was refreshingly cool in the Northeast, what would have been ideal conditions for exploiting the June air—were it not for the smoke. In the atmosphere, nothing happens in isolation.

SMOKE SIGNALS

Clearing away the smoke, the evidence strongly suggests that the atmosphere is sending us smoke signals.

The NOAA declared that for the Northern Hemisphere, 2023 was the warmest meteorological summer—the June 1 to August 31 period in the meteorological community—on record.[37] The Southwestern United States endured a brutally blistering heat; in records dating to 1979, Arizona, Colorado, New Mexico, Nevada, and Texas all set new standards for heat-related deaths in 2023.

THE OUTLOOK

Can about 50 million computer models be wrong? Barring a catastrophic volcanic eruption that veils the planet with cooling dust or some unimagined cataclysm, the warming trend will continue, making killer heat waves more likely.

When they do happen, they are likely to target the most vulnerable, and the heat will be at its worst where those people live—the cities.

The following are tips from ready.gov for dealing with extreme heat:

- Stay hydrated. This cannot be emphasized enough.

- Wear loose, lightweight, light-colored clothing, and when outside opt for all the shade you can find.
- Wear a hat wide enough to protect your face.
- Look in on family members and elderly neighbors.[38]

CHAPTER 5

LIVING ON THE ISLAND

- The world's cities provide indisputable evidence that human beings definitely can alter climate in ways that have profound effects on the human body.
- Paved surfaces and buildings and the defoliating of the landscapes have turned them into low-grade hot plates that can become especially dangerous during heat waves.
- One of the grander research projects in the history of meteorology quantified what happens to the atmosphere when development accelerates.

+ + +

The cities of the world are grand manifestations of human ingenuity and creativity, the ultimate work of civilization in progress. They constitute "the most all-embracing, diversified, but also, of necessity, the most constantly changing work, or work of art, that the human aggregate undertakes," in the words of Austrian planner Roland Rainer.[1]

I know from long experience with urban life that a city also can be a dangerous place, an edgy reality long acknowledged and oftentimes exaggerated and exploited in literature and mass media. Only in recent decades have government officials and health experts and, yes, journalists, come to recognize that the city can become a frighteningly deadly place for reasons that have nothing to do with its real and imagined dark sides.

It has everything to do with something not usually associated with urban life—warmth.

To the extent the builders of cities have tamed nature, they also have unmistakably perturbed it. That was by no means part of any plan, and that has become a problem.

IT TOOK AWHILE

The understanding of how cities have affected the overlying atmosphere would come thousands of years after people first began living in them.

The efforts of urban planners, environmental scientists, architects, and meteorologists have yielded quantum improvements in the understanding of city–atmosphere interactions. But so much of what we know today resulted from the astute and meticulous observations of a pharmacist, a Benedictine monk, and a German immigrant to the United States who undertook one of the most fascinating experiments in the history of climatology.

Long before climate change became a daily source of international anxieties—long before the Industrial Revolution, for that matter—it was known and well-documented that human activity unintentionally could alter the behavior of the weather. By radically changing landscapes, humans have been imposing a form of climate change on local environments.

Quantifying those effects was another matter. A groundbreaking contribution came from the British pharmacist, Luke Howard. With extensive temperature measurements, Howard demonstrated conclusively that the climate of London was demonstrably different from that of areas outside the metropolis, and that this wasn't a natural phenomenon. "The temperature of the city," he wrote, "partakes too much of an artificial warmth."[2]

Howard first published his findings in 1818, with significant updates in a three-volume set 15 years later. In retrospect, it all seems so intuitive: His explanation for the differences was almost stunningly obvious. The countryside, he reasoned, represented a "plane surface, which relates openly to the sky." Breezes can travel freely. It also is a "storehouse" for moisture that is available for evaporation, which gives off that cooling effect like that of sweat vanishing from the skin.

In the city, the rain is "very speedily exhausted," buildings absorb heat, and "greatly impede" the flow of winds.[3] In what would be a prescient insight, Howard concluded that given the impacts on solar radiation, evaporation, and winds, "We shall perceive that a city like London ought to be more heated by the summer sun than the country around it."[4]

In short, cities were becoming what these days we might call low-grade hot plates.

A GROWTH SPURT

Although no one was tracking temperatures back then—hard to do without thermometers—in all likelihood, urban warming dates to the days when humans first started congregating in cities. Ancient Rome almost certainly was warmer than the unbuilt outskirts, as was the case with Babylon 1,000 years earlier, when it had 1.2 million inhabitants.[5] What is different today is the proliferation of these urbanized hot plates, and the stunning increases in their densities.

Neither Luke Howard nor anyone else alive when his opus, *The Climate of London*, was first published, could have envisioned the dimensions of the urban explosion to come.

Consider what happened in just the United States. In 1800, the United States had only one city with a population of 100,000; by 1951, it had more than 190.[6] In fact, in the early nineteenth century, the four largest U.S. cities had a total population of 180,000, and no other cities had populations greater than 10,000.[7]

YOU DON'T NEED A WEATHERMAN . . .

As far as anyone has determined, Howard was the first to identify and quantify the effects of urbanization on temperature, a phenomenon now widely recognized as a major contributor to heat-related mortality. It was a historic achievement, but actually Howard had gained celebrity long before the appearance of the first volume of his London treatise. His fame was a testament to his curiosity and perspicacity.

In 1803, a 32-page, self-published pamphlet on the classification of cloud systems—his "cirrus," "stratus," and "cumulus" categories are still used today—became wildly popular. Its most prominent champion was the great Johann Goethe, one of Germany's most revered writers, who also became a well-regarded scientist. He lionized Howard's insights in poems about the clouds. So taken was he by Howard's work that the elderly Goethe sought to get in touch with the British master of the clouds, who at the time was no more than 21.

Goethe sent him a letter that was so flattering that Howard suspected it was a hoax, a prank concocted by his friends. Eventually he was persuaded that the letter was real, and in response sent Goethe a short autobiography.

Goethe's reaction must have overwhelmed the youthful Howard. "Nothing has given me so much pleasure as the autobiography of Mr. Howard," Goethe wrote. "Nothing more pleasant could have happened to me than to see the tender religious soul of such an excellent man opened out to me."[8]

Howard's resume was so impressive that he often has been called "the father of meteorology."[9] He was not a trained meteorologist, however.[10]

Perhaps it took a nonmeteorologist to name the clouds and divine something so evident—that built-up population centers would be hotter than the green open spaces. How significant Howard's work would become.

I learned about Howard's urban heat island research from another nonmeteorologist, a Benedictine monk, the Reverend P. Albert Kratzer. The first edition of his comprehensive tome, *The Climate of Cities*, was published in 1937 and updated in 1956. It was a global tour of urban-warming effects. Kratzer introduced a term that I particularly liked, "mesoclimate," which he called "a climate for human beings within a limited area." Rather than a "microclimate" effect that bears "no relation to its neighbors," he argued that "on the contrary, it bears many such relations. As an example, influences of the city on large-area climate and on the course of the weather in general are completely within the realm of possibility."[11]

Kratzer was on to something. Modern research by the University of Georgia's J. Marshall Shepherd and others has demonstrated conclusively that, indeed, urban heat has effects on rainfall downwind of a city.[12]

THE NAMING

The recognition and exploration of what would become a deadly urban warming trend was slow to evolve in the United States, but attracted considerable interest in Europe. In 1956, Father Kratzer, who in addition to being a monk was a geography teacher at the University of Munich, published an even more comprehensive survey of what was known about the urban environment.

It was two years later—and 140 years after the first publication of Luke Howard's book—that the city warming phenomenon got a name, the "urban heat island" effect. The legendary British climatologist Gordon Manley used it in a 1958 paper—not about hot weather, by the way, but about snowfall patterns. He hypothesized that urban warming might depress snowfall totals in the city, but probably not by much.[13] That was a bit off the mark.

Perhaps not surprisingly, in the United States interest in the urban climate was fired by a German immigrant, Helmut Landsberg, who was to become a towering figure in biometeorology. Landsberg, who received his doctorate from the University of Frankfurt, came to America in 1934 to teach classes in geophysics and meteorology at Penn State. His seminar on bioclimatology became the nation's first such graduate course. In 1941, he moved on to the University of Chicago, an international center of atmospheric research, and in 1967 he was appointed to a position at the University of Maryland.[14]

For the future of urban heat island research, this was a fortuitous assignment.

THE COLUMBIA EXPERIMENTS

In that same year, 20 miles from the Maryland campus, a grand experiment in urban design officially began on what had been about 15,000 acres of farmland and open space. The investor who had purchased the land was James Rouse, a star in the real estate universe. Rouse, a pioneer in the development of enclosed malls, said he wanted to build a community that would be "a garden for the growing of people."[15] Evoking William Penn's "Holy Experiment," Rouse envisioned "making a city into neighborhoods where a man, his wife and family can live and work and, above all else . . . grow in character, in personality, in love of God and neighbor and in the capacity for joyous living." This city, he said "will be economically diverse, poly-cultural, multi-faith and inter-racial."[16]

Columbia actually never quite became a city, and remains a "Census-designated place," governed by Howard County. Its ancestry dated to the opening of the "Columbia" post office in 1874. Columbia was located at the crossroads of Columbia Road, Route 29 these days, and Old Annapolis Road, which today is Route 108. As late as 1912, the post office served all of 20 residents.[17] James Rouse's redefined version of Columbia officially was dedicated on June 21, 1967, coincidentally the day of the summer solstice.

In Rouse's grand urban experiment, Landsberg saw an opportunity for a grand science experiment in how urbanization can change the weather. It began a year later, in what Landsberg described as a place that was "no more than a crossroads," and "partly wooded and slightly rolling," with only 200 residents.[18] It was perfect for what he had in mind. Rouse saw a "garden for the growing of people," but Landsberg saw a garden for the growing of a heat island.

"CRUEL EXPERIMENT"

Evidence of urban heating emerged grimly during World War II. Available observations showed that the decimation of cities by bombing had cooling effects on the local environments, in what Landsberg called "a cruel experiment." However, the period of record was "too skimpy for scientific assessment." In this case, that lack of data was the result of a positive development for human beings: "Fortunately," Landsberg observed, "reconstruction soon made cities rise again from the ashes."[19]

Columbia, by contrast, was rising from the ground, so much of which it would be paved over. Thus, it afforded the perfect opportunity to chronicle the ripening of a heat island as the development built its way toward accommodating 100,000 residents.

Landsberg reached an agreement with the Rouse Company that permitted his team to set up instrument arrays that monitored temperatures and other conditions during various stages of construction. The study covered a seven-year period, from 1968 to 1975, when the population grew from 200 to 20,000, and construction proceeded vigorously.

The study affirmed and verified that cities are indeed low-grade hot plates and "confirmed in most respects earlier deductions," Landsberg wrote in his paper, published in 1979.[20] He identified the trend that health experts later would target as being particularly hazardous for the vulnerable: "An unmistakable urban heat island has developed that is quite notable nocturnally." The biggest increases in temperatures occurred in the areas of the shopping mall and the town center.

The new Columbia effectively erased "a rich fabric of microclimates," created by the mix of stream valleys, woods, hills, and flora. (In retrospect, Landsberg said he regretted not including scientists to study impacts on flora and fauna.) What supplanted that fabric was a microclimate "governed by density of buildup." The escalating warmth resulted from the absence of vegetation, which could have supplied sources of moisture for evaporative cooling, and the "high heat absorption of the building materials." Urbanized surfaces can significantly decrease sun-reflecting albedo, soaking up solar energy and only begrudgingly giving it up.[21]

This was all from the Luke Howard playbook.

HOT NEIGHBORHOODS

Averaged over the entire urban environment, the Environmental Protection Agency (EPA) estimates that daytime temperatures in cities are 1 to 7 degrees Fahrenheit higher, and 2 to 5 degrees Fahrenheit warmer at night than outlying areas.[22] Those differences are more than significant and merit attention and concern; however, they do not speak to the dimensions of the heat hazards.

The nighttime departures are especially troubling. Government heat advisories historically have warned against overexertion in daytime and underrated the dangers attending the lack of nighttime cooling. Plus, those citywide averages do not speak to the reality that in the end, all weather is local. The effects of urban heating vary from neighborhood to neighborhood; as the Reverend Kratzer observed, "Every street has its own climate."[23]

"Hot neighborhood" has taken on a whole different and haunting meaning. One study by the advocacy group Climate Central estimated that two-thirds of New Yorkers were living in neighborhoods with temperatures 9 degrees Fahrenheit higher than those of nonurban areas.[24]

Dissecting the 2003 heat-wave deaths in Paris using satellite imagery to discern temperature patterns, a team of French researchers concluded that along with elevated temperatures in the vicinity of buildings, warm overnights were killers. "In elderly individuals, exposure to a high nighttime temperature over several days increases the probability of death during a heat wave in urban conditions, whereas daytime temperature is less important," they said.[25] The most vulnerable were those suffering chronic diseases, and with impaired mobility, another team found.[26]

In retrospect, the 1995 tragedy in Chicago foreshadowed what was to happen in Europe. The most vulnerable in Chicago included those with pre-existing medical conditions who were confined to bed. Most of the victims were elderly.[27] And most of the Chicago victims had lived "in the heart of the urban area." Some were hesitant to open windows, perhaps for fear of crime. During the heat waves of the 1930s, "many residents slept outside in the parks or along the shore of Lake Michigan."[28] So many victims were elderly people who live alone in the warmest neighborhoods. The ranks of the endangered population are likely to swell along with the increase in the live-alone population.

COOLING THE ISLANDS

The reality of the heat hazard has been made all the more evident with those changes in how heat deaths are calculated. Relying on excess mortality, rather than ascertaining core body temperature at the time of death, has validated the argument for heat being the planet's number-one killer.

Only in recent years have health officials come to realize how truly dangerous it is to be poor, elderly, and alone in a city neighborhood. Unmistakably, localized manmade climate modifications are having deadly consequences, with an additional contribution from worldwide warming. As global temperatures rise, the water-vapor content of the atmosphere also rises for the simple reason that warmer air can hold more moisture.[29]

All that moisture inhibits radiational cooling in cities at night.[30]

Cities have become smarter and savvier about the manmade hazards and have been taking measures to turn down the heat, including commonsense measures such as planting trees and introducing more vegetation. Temperatures in a grove of shade trees can be as much as 9 degrees Fahrenheit cooler than surrounding areas. Readings in a suburb with mature trees can be 4 to 6 degrees lower than those in a treeless new development, according to the EPA.[31]

LESSONS UNLEARNED

Tens of thousands dead in Europe in 2003. Several hundred dead in Chicago in 1995. Hundreds of thousands, perhaps millions of the uncounted throughout the world have died as a result of heat given an extra jolt by the urbanized environment. Could all those people have survived the hot spells?

I would argue that heat-related death tolls would have been greatly reduced if government officials and the medical community had been aware of, and had acted upon, Helmut Landsberg's insights and observations from his great Columbia experiment of a half-century ago.

One study conducted in 93 European cities in the summer of 2015 found that urban warming was a contributing factor in 95 percent of all deaths. The researchers' findings about the lifesaving potential of urban vegetation[32] suggests strongly to me that heat-death totals worldwide could have been greatly reduced if decision-makers had paid attention to Helmut Landsberg's work.

Landsberg ended his analysis of the Columbia experiment with a rather stinging critique of some of the environmental aspects of the Rouse project. He noted that bulldozers had claimed "a large number" of mature trees that could have provided heat-mitigating shade. They also would have retained moisture that would be available for diverting some of the sun's power to evaporation. The Columbia area had many old trees, but a large number of them succumbed to the bulldozer to make construction easier. Only Symphony Woods and a few isolated old trees survived in the area. He argued that Rouse would have been better off saving some trees and creating parkland rather than artificial lakes. While they may have been a pleasant aesthetic addition and recreational assets, those bodies of water were too small "to have an appreciable effect" on temperatures. He estimated they cooled areas only within about 65 yards of the shores.[33]

Landsberg thought the parking lots constituted a bad idea, that it was better to opt for enclosed garages that would reduce the overall ground cover. He concluded, "Obviously, as far as heat island development, . . . Columbia showed no significant advances over other towns of comparable size." Summarizing his findings, he concluded that Columbia's resistance to the remediation of heat island effects was "inexcusable." "There is no dearth of information on urban micro- and meso-meteorological conditions, but this has been largely ignored in planning."[34]

And based on what we know now about heat-related mortality in the cities, the consequences at times have been tragic.

The heat-island effects have been dangerously localized, down to the block levels of neighborhoods. It would be impossible to know how many heat-related deaths have occurred in urban environments through the centuries. However, those well-documented heat disasters in Chicago and Paris have affirmed that heat is the number-one weather-related killer.

CHAPTER 6

"DYING ALONE"

- Heat-related tragedies in the United States foreshadowed what may have been the worst weather disaster in modern history.

- Extreme heat, exacerbated by the urban heat island effect, led to tragedies that ironically could have happened only in the developed world.

- The overarching concern among weather-mortality and climate experts is whether disasters in Chicago and Europe were chilling previews.

+ + +

It did not involve a hurricane, tsunami, tornado, or blizzard, no images of homes ripped from their foundations or yachts piled upon beaches or waters coursing through streets or trains buried in snow.

One of the most horrific weather tragedies in human history occurred in slow motion in the summer of 2003.

"Occurred" probably is the wrong verb. "Unfolded" might be more appropriate, the enormity of it all obscured until it was far too late to stop it. Tens of thousands, and maybe as many as 70,000, died in Europe as a result of insufferable heat that ripened during that July and turned deadly that August.[1]

How could this have happened? As it turned out, it was a disaster that could have happened only in the developed world. It was foreshadowed by heat wave tragedies—not in Bangladesh or Somalia—but in Chicago and Philadelphia.

It could have happened only in places where progress in medical science and improved standards of living had increased life spans, while at the same time the ranks of the forsaken population were swelling. So many of the victims were vulnerable to all the ways that extreme heat could compromise the body, and so many of them were left to die alone.

It was at once a natural and unnatural disaster, the culmination of a seminal 10-year period in which disasters in Philadelphia, Chicago, and throughout Western Europe conspired to raise international awareness of the dangers of heat. Perhaps the worldwide reactions and responses contributed to the fact that, while the planet indisputably warmed, heat-wave deaths would decline in the next two decades.

Whether that trend will continue is another matter. Eric Klinenberg, a prominent sociologist who wrote a comprehensive and chilling analysis of Chicago's 1995 heat wave, is among those who are confident that it won't. The recent trend, he says, "is just a matter of luck."[2]

Deadly heat in Europe in 2022,[3] and the historic heat wave in the U.S. Southwest in the summer of 2023 suggested that luck may well be on thin ice.

THE "AUTOPSY"

Klinenberg is haunted by the 2003 death toll and what it might mean to the future of New York, Detroit, Philadelphia, his native Chicago, and urban centers around the world. He still wonders if he could have taken some action that would have mitigated the tragedy in Europe.[4]

That year, Europe experienced its hottest summer since the Renaissance. Based on satellite data, summer temperatures in Paris averaged 4.5 degrees Fahrenheit above normal, according to NASA scientist Dr. John Christy.[5] That's a stunning departure for summer, when heat is spread more evenly across the hemisphere, as opposed to the jumpiness of the more volatile cool seasons. For example, the warmest summer on record in Chicago was only 3.3 degrees warmer than the long-term average.[6] By the time those 2003 hot spells were over, the heat-related death toll was beyond anything the world had experienced.

Eric Klinenberg saw it coming. Evidently no one in a position to do anything about it, actually did. Klinenberg, who would become one of the nation's most respected and visible sociologists, had written a book about his native Chicago that was published the summer before.

As the title suggested, *Heat Wave: A Social Autopsy of Disaster in Chicago*, was not beach reading. It was an intensely reported dissection of the 1995 Chicago hot spell that was blamed for the deaths of at least 739 people. For my money, it remains the most important treatment written about heat wave hazards and the people most likely to be their victims.

"They target, poor, old, sick Black people," Klinenberg told me in August 2023. "We have a tendency to treat them as expendable." Hundreds of the victims "died alone, behind locked doors and sealed windows, out of contact with friends, family, and neighbors, unassisted by public agencies or community groups," he would recall in an interview posted by the University of Chicago Press, coinciding with the book's publication in 2002.[7]

What happened in Chicago all but foretold what would happen the next year in Europe, an event explored in detail in Richard C. Keller's impressively documented post-mortem, *Social Isolation: The Devastating Paris Heat Wave of 2003*, published in 2015. Had health officials in the European Union—particularly in Paris and in Italy—read Klinenberg's book or had even a PowerPoint–level familiarity with its contents, could tens of thousands of lives have been saved?

Maybe.

"Don't think I didn't get upset about that," he said. "I wish I had done a better job of getting the word out, outside the United States."[8]

CHICAGO FIRE

Ironically, Klinenberg was living in Paris in the summer of 1995, the year of the Chicago disaster. He was a 24-year-old preparing to enter Berkeley's doctoral program in sociology. He read about his hometown disaster in an English-language newspaper, the International Herald Tribune.

Five years later he would undertake his own investigative reporting project and write an account considerably more exhaustive.[9] Poring through the records, what he found was to him almost unimaginable. "It was gruesome and incredible for this to be happening in the middle of a modern American city," he would recall in an interview posted by the University of Chicago Press, coinciding with the publication of his book in 2002. In retrospect, so much of what he said was a harbinger of the horror to come a year later.

During that 1995 hot spell, it was impossible for paramedics to respond to all the emergency calls. The volume was so enormous that 23 hospitals stopped accepting patients in their emergency wards. Ambulance crews had to ride for miles shopping for openings.

Not surprisingly, hundreds of the victims never made it to hospitals for treatment. "The most overcrowded place in the city was the Cook County Medical Examiner's Office, where police transported hundreds of bodies for autopsies," Klinenberg said. By the third day of the heat wave, the morgue's capacity "was exceeded by hundreds." Cook County dispatched refrigerator trucks to store the bodies.

One perhaps forgotten element of the Chicago heat wave was the fact that the death toll literally was unbelievable to so many people in position to make decisions. The skeptics included Mayor Richard M. Daley, son and successor to the legendary Richard J. Daley.

A key figure in the drama was Dr. Edmund R. Donoghue, who had become chief medical examiner of Cook County in 1993. Coincidentally, that was the year of a deadly heat wave in Philadelphia that had a lot to do with how the world learned about Chicago's death toll.[10]

Donoghue's counterpart, the late Dr. Haresh Mirchandani, was the chief medical examiner in the City of Brotherly Love during what was a punishing summer in the Northeast corridor.

REVOLUTION IN PHILLY

In New York, the temperature reached 100 degrees Fahrenheit on three consecutive days, July 8, 9, and 10, 1993. Baltimore made it to 100 degrees twice and 99 once, and Philadelphia, 100 on all three of those days.[11] What was different about Philadelphia was the fact that the heat was killing people, lots of them, at least according to the city's

Health Department. By contrast, heat deaths were scarce in New York and Baltimore.[12]

Philadelphia was feeling the heat in more ways than one.

Ed Rendell, who later would become governor and chairman of the National Democratic Committee, had enough on his plate in weathering a mammoth fiscal crisis. He would weather his share of Mother Nature's storms during his eight years in office, in a decade marked by extremes all over the nation. For anomalous behavior, during the 1990s the percent of the nation that experienced the atmosphere's extremes was even greater than it was during the Dust Bowl era of 1930s, according to the National Centers for Environmental Information.[13]

Rendell had taken office in January 1992 and had led the city through the March 1993 "Blizzard of the Century." This was his administration's first experience with a serious heat wave. The official death toll raised questions about the city's response, recalled David L. Cohen, who was the mayor's chief of staff at the time and now is the U.S. ambassador to Canada. If it could let all these old and poor people die while they appeared safe in other cities, Philadelphia certainly wasn't presenting itself as the "City of Elderly Love."[14]

On another front, the administration was undergoing a credibility crisis. The death toll was met with howls of skepticism about how the city was keeping score. Perhaps the most stinging came from Detroit, where Mirchandani had served before moving to Philadelphia. Bader Cassin, the Wayne County, Michigan, medical examiner, ripped the Philadelphia system for calculating the heat-related death numbers: "It's a guess. It's a really sloppy statistic." As evidence, he pointed to the fact that people weren't dying in other cities, including Detroit. Cassin said that Detroit, "had temperatures in the 90s for over a week, and we haven't had more than one or two deaths caused by the heat."[15]

Rendell and Cohen were under pressure and wanted answers.[16] What they didn't realize at the time was that Mirchandani had started a revolution in the manner in which heat-wave deaths were calculated.

The traditional criterion for a heat death was the establishing of hyperthermia, a core body temperature of 105 degrees Fahrenheit, verified by an investigator. Mirchandani had expanded the definition to include deaths in which heat was identified as a contributing cause.

His reasoning? In a prolonged heat wave with ultra-hot nights, the doctors on his skeletal staff would never be able to get to all the victims in time to confirm hyperthermia at the instant of death. He told his staffers to consider forensic evidence, such as windows nailed shut, or a fan circulating the hot air in the room. If they could establish that heat contributed to a fatality, it would be counted as a heat death.

Philadelphia's counting system served to "sound the alarm" that the heat can kill. "It is not a false alarm," said Mirchandani. "On a typical day, we see 15 cases . . . and if it suddenly jumps to 40 cases when the temperature is 100 degrees and we go out and see all the windows shut—it's a commonsense attitude."[17]

The Philadelphia counting method became a federal case. Three Centers for Disease Control and Prevention (CDC) investigators bearing the title "epidemic intelligence service officer" came to Philadelphia on August 12 for an intensive examination of the city's death records.[18] What would they find?

The city waited months for an answer. The suspense finally ended on June 14, 1994, when the CDC affirmed the validity of the Philadelphia method, and then some. The CDC's Sherrilyn Wainwright said that while the agency couldn't force other medical examiners around the country to follow Philadelphia's system of calculating heat deaths, it affirmed that the city was doing it right.[19]

In fact, it was found that similar increases in deaths during the heat wave had occurred in other urban areas near Philadelphia, although they had not been recognized at the time, according to a federal disaster report issued in December 1995.[20] From July 12 through August 17, 1995, 72 deaths in Philadelphia were attributed to the heat.

Echoing Mirchandani's reasoning, Wainright said: "We must consider a heat wave a public health emergency."[21]

In Chicago, Edmund Donoghue evidently embraced that message, and in 1995 he sounded the alarm as the deaths accumulated, withstanding the blowback from the mayor. Unlike Ed Rendell in Philadelphia, who responded to the skepticism generated by the City of Brotherly Love's

1993 heat-related mortality figures by accepting the medical examiner's reasoning, Daley publicly questioned Donoghue. He suggested that by Donoghue's reckoning, every death that occurred in summer could be considered heat-related. Some Chicago-area journalists appeared to share Daley's perspective as to whether the deaths were "really real."[22]

The CDC's affirmation of the Philadelphia heat-related death to Philly's system "really helped me," said Donoghue.[23] And the Chicago body count lent further credibility to Mirchandani's methodology, said Dean Iovino, a lead meteorologist in the weather service's Philadelphia-area office, who was on duty during that heat wave and well remembered the scoffing at the death counts.[24]

Medical examiners across the country stood behind Donoghue, who reported 521 deaths for the month of July. As it turned out, Donoghue was wrong, although his counts were far superior to Mayor Daley's estimates. When the final figures were tabulated for "excess death" numbers—the difference between reported deaths and the average mortality for a given period—deaths exceeded the averages by 739 for the week beginning July 14. In other words, "Donoghue had been conservative in his accounts."[25]

Donoghue said he was well aware that he would have to wait for vindication, that collecting and evaluating the excess mortality figures would be a laborious process. "It's going to be a little while before the cavalry came to save you," he said. But he didn't flinch, and three decades later he told me, "I wouldn't do anything differently. We did the correct thing."[26]

What conditions caused such catastrophic loss of life?

CATASTROPHE IN EUROPE

The meteorological antecedent conditions were ideal for the summer of 2003 to turn hellish and deadly in July and August in Europe.

Months of below-normal precipitation left the soils parched, ripe for baking by the sun. The Danube River reached its lowest levels in over 100 years. Fish and avian populations were drastically reduced. The Mediterranean Sea surface temperatures off the coast of Spain approached 90 degrees, the warmest ever recorded.

Wildfires were widespread, and they may have had impacts on the overall dryness. For rain, warm air has to rise over cooler air, and the aerosols may have inhibited that process.

A persistent, late-summer dome of high pressure smothered the continent during the crucial heat wave period, at once repelling rain, keeping the soils dry, and enabling the heat to build.[27]

Under high pressure, the air is heavier as currents descend and dry out the atmosphere. Conversely, contrary to popular perception, the air isn't "heavy" during storms. Storms are associated with low pressure, or lighter air: Parcels of air have to ascend for water vapor to condense into rain. Low pressure had no chance as the heat built up that summer. Dryness begets dryness, and allows heat to beget heat. The daytime sun didn't have to waste energy evaporating moisture from foliage or other surfaces.

The French were slow to recognize that a tragedy was developing, said Richard Keller, whose book *Social Isolation* was published in 2015. As the heat was building, it was more or less a feature story. Pictures of people splashing in fountains were common on newspaper front pages. "For the most part the heat was billed as an inconvenience," said Keller, a professor of medical history and population health at the University of Wisconsin-Madison. France, where an estimated 15,000 people died from heat-related causes, appeared to be the nexus of the disaster. France actually had experienced three heat waves that summer, with the one in August being "the whopper," according to Keller.[28]

The bulk of the fatalities occurred in August in urbanized areas, especially in and around Paris. Based on excess mortality figures, 300 had died by August 4.[29] The French meteorological agency sounded the alarm, but underscoring a point that Klinenberg made about traditional media coverage of heat waves, newspaper front pages showed images of people splashing in fountains. More dangerously, the government was slow to admit that a crisis was building, despite a warning from the head of the union for emergency room doctors.[30]

The heat death toll rose astronomically, jumping to 3,900 by August 8.[31]

The following day, Health Minister Jean-François Mattéi addressed the crisis—from his vacation home in the Riviera. He was wearing a polo shirt. He assured the audience that France was on the case, but that estimating just how many were dying from heat-related causes was not a cut-and-dried matter. The deaths, he said, "could be related to heat, but you know there is really an intricacy of the phenomena. These are often patients who have a chronic illness, they are people who are fragile, at the end of their lives," he said.

By August 10, more than 6,800 people had died of heat-related causes. "At this point, the bodies were just absolutely piling up in make-shift morgues in cities like Paris and Lyon," Keller recalled.

The August toll would soar past 14,800.[32]

More than a third of the victims would die in their residences. The intensity and duration of the heat clearly was overwhelming the bodies of the vulnerable, compounded by the fact that overnight "lows" weren't falling much below 80.[33]

Historically, so much of the language of heat warnings has emphasized limiting outdoor activity on hot days. But numerous studies—including one in May 2023 by Japanese researchers, who used overnight lows of 77 or warmer as criteria and looked at nearly 25 million deaths—have documented that hot nights are killers since they offer no respite for the vulnerable.[34] As we've mentioned, thermal comfort is relative.[35]

It is also possible that "an extremely high level of ozone" may have contributed to mortality.[36]

ABOUT OZONE

Unfortunately, ground-level ozone—not to be confused with the "good ozone" in the upper atmosphere—is a staple of summer in the cities. It is toxic mix brewed at the surface when pollutants emitted into the air from vehicles, power plants, refineries, and the like interact with "volatile organic compounds." It is the summer sun that lights the stove. Ground-level ozone is most likely to exceed "unhealthy levels" in urban areas when it's hot and sunny.[37]

And that is a mild description of what was going on in Europe in the summer of 2003. The combination of a lack of cloud cover and heat were unrelenting, creating a hazardous harvest of ozone in parts of Europe. "The situation was particularly favorable for significant ozone formation for several weeks in south and central parts of Europe," observed a group of Norwegian researchers.[38]

IT HAPPENS IN ISOLATION

Indisputably, however, social isolation was an overpowering contributor to the massive mortality figures. Among those left alone to die were people with serious background conditions, such as heart disease and perhaps psychiatric issues. In some cases, the victims might have been using drugs that inhibited thermoregulation.[39]

In its post-mortem, the French National Assembly noted that, along with the quality of residential construction, the wholesale abandonment of the vulnerable population were two overarching factors in the catastrophe. On the issue of construction, the Assembly said that a "large majority" of buildings inhabited by the victims had zinc roofs, "a metal transmitting heat all the more easily as the thermal insulation of the ceilings of apartments on the top floors of buildings is often insufficient, or even nonexistent."[40] They were the French equivalent of the U.S. brick rowhomes that Dr. Lawrence Robinson, formerly of the Philadelphia Medical Examiner's Office, said could turn into deadly "brick ovens" during a heat wave. Robinson told me that he had seen heat melt unburned candles. And if the houses don't have a chance to cool down at night, when the sun comes up they get hotter in a hurry.[41] But the French Assembly concluded that the death toll would have been so much lower were it not for "the isolation of many elderly people in large towns."[42]

This was a source of national shame.

Those with the means and the opportunity sought asylum from the heat away from the urban centers. They returned to the horrors of the nightmare that had haunted France while they were away. "Some found gruesome scenes: apartment buildings pervaded by the reek of decomposing flesh; ceilings stained by bodily fluids that had dripped through from the flat above, where an upstairs neighbor had died and lain undiscovered for weeks."[43]

It was all so eerily similar to what had happened in July 1995 in Chicago, when bodies piled up in the city morgue as heat deaths soared past the city's ability to deal with the corpses.[44] As it happened, 1995 was the same summer that Philadelphia developed its heat-response plan with the aid of international heat-mortality expert Dr. Laurence Kalkstein, which became a model for other U.S. cities. A study of mortality in the three summers after the plan went into effect estimated that it had saved 117 lives in Philadelphia.[45]

In the summer of 2004, France instituted a variant of that system. It involved a heat-warning system in which meteorologists and health specialists advised local officials on when to sound the alarm bells and take such actions as checking on the homeless. It also enlisted what may be the ultimate vaccine against heat-related dangers—air-conditioning.

THE LIVES SAVER

As in the United States, when the heat was on, the French opened "cooling centers" in public spaces to offer potentially life-saving relief to those who didn't have access to air-conditioning.[46]

Air-conditioning comes with an environmental price, and with manufacturers targeting markets in Asia and Africa, it is unclear how the world will respond to a rapidly growing electricity demand. But it would be impossible to calculate just how many lives have been saved through the decades by this cooling revolution, and how many would have been spared in Philadelphia, Chicago, and Europe if more of the vulnerable would have had access to air-conditioning or had found their ways to cooling centers.

One group of researchers concluded that deaths from extreme heat plummeted 80 percent from 1960 through 2004, compared with the 1900–1959 period. They argued that the drop was almost 100 percent due to air-conditioning.[47]

In retrospect, it is hard to imagine that only the tiniest percentage of Americans had air-conditioning before 1950.

ADDENDUM
THE COOL REVOLUTION

While Helmut Landsberg was tracking the making of a heat island in the late 1960s and 1970s, an all-out revolution was underway, unrelated to the civil rights battles and the anti–Vietnam War tumult. This was a revolution wrought by technology, born of centuries of trial and error, a physician's efforts to conquer malaria, a humor magazine's dilemma, and the ingenuity of young engineers.

The air-conditioning of America was underway, in a cold war that the nation would win resoundingly.

Air-conditioning has indisputably saved lives and became a prime agent for massive changes in U.S. demographics. It has instituted an era of climate change at the most local level, where the majority of the population spends the bulk of its time on hot summer days and nights—indoors.

It can help people sleep. It can make them smarter.

It has dark sides. It can make some people sick and, ironically, the grand cooling movement might be contributing to the warming of the planet, with its massive consumption of electricity and the use of hydro-fluorocarbon coolants in older systems.[1]

Air-conditioning unquestionably changed society's relationship with summer, and to say that it has transformed America is no exaggeration—its transformative power in a league with the railroad and the lightbulb.

So many homes and buildings in the United States are air-conditioned these days, it may be hard to imagine what the nation was like without it. Yet, its ubiquity is a relatively modern development.

THE BEGINNINGS

Willis Carrier's role was seminal and has so much to do with how air-conditioning catapulted from luxury status to necessity and ubiquity. The units bearing his name remain near the top of the lists that rank the business elite in terms of quality and sales.

But the history of air-conditioning in America more properly begins with a doctor who saw it is a prime weapon, not so much against heat, but malaria—and who evidently never actually built a cooling system.[2] And save for the avarice of business interests who saw him as a threat to their profits, that physician, Dr. John Gorrie, might well have been at least as famous as Willis Carrier.

Details of Gorrie's life and career vary, depending on the sources. It is known that, like so many eighteenth- and nineteenth-century figures, Gorrie had eclectic interests, among them how to coexist with extreme heat. That made sense, as he had settled in the heart of the Florida Panhandle, in the torrid, steamy city of Apalachicola, in 1833. Those who have pieced together details of his life agree that he was heavily involved in the town's civic affairs and later became mayor.[3]

In addition, Gorrie was an attending physician at a local hospital. He became well-acquainted with the ravages of malaria, an affliction whose impacts on the world would be hard to overstate. Its connection to the Anopheles mosquito wasn't established until 1897, but malaria's existence has been traced to ancient Egypt. It was spread during the Roman conquests, and some theorize that it contributed to the fall of the empire. It was blamed for killing up to 300 million people in the twentieth century alone, and who knows how many people before then.[4]

Dr. Gorrie lasered his attention on malaria, and while he obviously was unaware of the mosquito connection—he died more than 40 years before it was made—he correctly concluded that the presence of the disease in the Apalachicola area was related to the hot, moist, and oppressive climate in summer and fall. Gorrie advocated vigorously for draining the swamplands in the area.[5]

In the summer of 1841, it was reported that 47 people died of malaria in Apalachicola alone, and that ghost ships bearing the bodies of deceased malaria victims were adrift.[6]

Gorrie became convinced that cool air would be the antidote, that "Nature would terminate the fevers by changing the seasons."[7] To treat malaria-infected sailors, Gorrie concocted an "air-conditioning system" in which air was blown over buckets of ice placed in the sickrooms. Just what powered the airflow was unclear.[8]

The concept was similar to a cooling method that Australians have recommended, and they do know a few things about cooling Down Under. "A cleverly positioned bowl of ice is all you need to turn a fan into a cold mist machine, . . . Place a shallow bowl or pan of ice in front of a fan for an icy-cool breeze that won't break the bank."[9] (I wish I had thought of that in my youth when our air-conditioning-deprived house was far too hot for sleeping.)

Another account of Gorrie's cooling method described a somewhat more elaborate, but still primitive, system. He filled an urn or other container with ice. This was suspended from a ceiling in such a way that outside air, transported by a pipe, flowed downward over the ice. Cold air sinks, and the air flowed outward through a ground-level pipe.[10] Again, however, it was unclear precisely what made the air move. Whatever the method, Gorrie's system was wanting for raw material: the Florida Panhandle was not exactly an ice capital.

In those days, ice arrived at Gulf ports from Boston and New York, mined from bodies of water in the North. It was an enormous and lucrative industry.[11] For a variety of reasons, however, the shipping schedules were undependable. The vessels had to navigate past Cape Hatteras, where the chilly air oozing off the continent interacts with the warm waters of the Gulf Stream to ignite perilous and destructive nor'easters. That region of the Mid-Atlantic coast is one of the most dangerous storm-breeding grounds in the world.[12] When it actually did arrive, the ice was pricey, and scarcity made it pricier.

Gorrie was determined to find a way to end his dependence on ice imports.[13] What he envisioned transcended passing air over ice, something beyond using this in hospital and sick rooms.

The combination of high temperatures and humidity "prevents a large portion of the human family from sharing the natural advantages they possess and causes mental and physical deterioration of the native inhabitants," Gorrie wrote in the *Apalachicola Commercial Advertiser* in 1844. In another article that same year, he described a plan for "an engine for ventilation, and cooling air in tropical climates by mechanical power" that "may be placed in any part of house or ship." This "compressed air refrigerating system" would make possible "a future when fruits, vegetable, and meats would be preserved in transit and thereby enjoyed by all."[14]

Although Gorrie never did execute that vision, his obsessive efforts built an important bridge in the journey toward the air-conditioning systems we take for granted today. Plus, his efforts yielded an immeasurable bonus in that he secured the first U.S. patent for a device that could chill the air and make ice.[15]

Unfortunately, Gorrie never profited from his invention, and his professional and financial life collapsed after he received the patent. His primary investor and chief source of funding died. Critical and vituperative articles in Northern newspapers chilled interest among other investors. Gorrie believed that Edward Tudor, who all but controlled the world ice market, exporting thousands of tons cut from frozen waterways to the South and all the way to Calcutta, was the force behind the discrediting of his invention. Gorrie died broke and in seclusion in 1855.[16]

Gorrie has been called "the father of air-conditioning," although that paternity would be hard to establish. As groundbreaking as his work was, Gorrie had built on the ingenuities of centuries of predecessors: Leonardo Da Vinci, for example, constructed a water-driven fan to cool the bedroom of his patron's mistress around 1500. While no individual could be credited with inventing this world-changing technology, it would not be a stretch to say that a twentieth-century engineer notorious for his absent-mindedness would become the father of modern air-conditioning. He got his start in one of the snowiest regions of the country, at a company created to serve a dying industry.

THE ROAD TO UBIQUITY

The Buffalo Forge Co., an enterprise founded as a manufacturer and supplier of blacksmith forges, hired a young Cornell University graduate named Willis Carrier in 1901 with the intention of having him join its sales force. The person attempting to train him, however, thought the new hire might be better suited to research and development.[17] It was a transfer that worked out well for both parties.

While the era of the blacksmith was entering its twilight, Buffalo Forge owners William and Henry Wendt were getting a jump on the future. Instead of traditional blacksmith bellows, the company was using "gear-driven blowers." Those forge fans helped propel the company into the air-movement business.

The company was becoming a pioneer in heating and cooling systems, circulating warm air through heated coils to produce heat and, reminiscent of John Gorrie, cooling areas with the movement of air over ice.[18]

One might conclude that Buffalo Forge and Willis Carrier were made for each other.

For my money, the term "absentminded" is a misnomer. So many reputedly absentminded people I've met in my lifetime were altogether present-minded. It happened that their minds were so laser-focused on a certain thing or things that they tended to struggle with details that others processed so effortlessly.

Carrier was described in a *Time* magazine profile as "absent-minded about meals and haircuts."[19] He once volunteered to make reservations for a group at a hotel restaurant, but by the time he got there, he had forgotten about his mission—not to mention the other members of his party. They were left waiting in the lobby while Carrier ate alone.[20]

The distraction issues notwithstanding, Willis Carrier turned out to be a pretty good hire. Unlike Gorrie, Carrier didn't set out to develop a cooling system that would heal the sick and make homes comfortable in the heat. Rather, what he accomplished had a lot to do with the plight of a popular humor magazine.

Having shown promise as a system designer at Buffalo Forge, working up drawings for the likes of a wood-fired kiln and even a coffee dryer, in 1902 he was assigned to work on a problem bedeviling the Sackett & Wilhelms printing company.[21] Among its customers was *Judge* magazine, started by defectors from its chief rival, *Puck*.[22] Moisture levels at the Brooklyn printing facility were the mortal enemies of the color presses. With changes in humidity the paper stock would swell and contract, and that would create problems with aligning the colored inks. The result was paper waste and lousy print quality, not to mention lost production time.

A Sackett & Wilhelms consulting engineer presented the problem to a Buffalo Forge official to gauge the company's interest in getting involved. The official said yes, and he had a specific engineer in mind to help solve it. The problem was placed fortuitously in the lap of Willis Carrier.

While engineers had figured out ways to modify the air, this was unique challenge. The mission was to find a way to control the moisture level of the indoor air consistently.

Misson accomplished: On July 17, 1902, Carrier placed his initials on mechanical drawings that addressed the printing company's issues.[23] His system controlled the moisture content of the air and the temperature by blowing air over refrigerated coils.[24] Carrier's system provided the company with "excellent results," and the company that today bears his name marks July 17, 1902, as the birthdate for "modern air conditioning."

But that was well before the minting of the term "air conditioning," and it wasn't Carrier who did the minting. That distinction belongs to one Stuart Cramer, a factory developer, and the naming happened almost incidentally. As was the case with Gorrie, Cramer's contributions to the technology were for the most part under the historical radar.

Cramer invoked the term in a speech he delivered in May 1906 at the American Cotton Manufacturers Association convention in Asheville, North Carolina. He was there to discuss the "Automatic Regulator" he had invented to control humidity levels in textile factories.[25] A decade earlier, Cramer, a U.S. Naval Academy graduate, introduced the process of "yarn conditioning," controlling the moisture content of the air inside textile mills.[26] Perhaps the seminal reference in the title of his talk to the cotton manufacturers—"Recent Development in Air Conditioning"— was an unconscious play on words. Cramer pointed to what he called the "interdependence" of temperature and humidity. "I would also mention air cleansing," he said. "And, so, I have used the term 'air conditioning.'"[27]

The term didn't make much of an impression on the journalists in attendance. Their news stories dealt mostly with the issue of factories poaching employees from each other. (And, gee, when was the last time a

journalist whiffed on a big story in the pursuit of something less important?) But it wasn't long before Carrier and his colleagues embraced the
phrase "air conditioning" and, eventually, so would the rest of the world.

Incidentally, Carrier recalled that he had met Cramer only once,
and that was on a train. "I found him to be a fine Southern gentleman,"
he said.[28]

In 1905, Carrier became the chief of the Buffalo Forge engineering
department.[29] His career would experience a quantum leap four years
later. The Buffalo Forge managers, having realized the opportunity they
had with this emerging technology, decided to create a wholly owned
subsidiary. Honoring the man whose work created this potential revenue bonanza, they named it the Carrier Air Conditioning Company of
America.

The customers lined up. Celluloid Company, which was producing film for the nascent motion picture industry, signed on as a client.
Textile companies followed, as did the Gillette Safety Razor Co. An
air-conditioning system mitigated rust on razor blades at the Gilette
Razor Co. plant. Other customers included rubber and baked-goods
factories, along with a hospital that used air-conditioning in its ward for
premature babies. A system installed at a dust-choked tobacco plant in
Richmond did such an efficient job of cleansing the air that workers were
able to eat lunch on the job.

Then the Great War interrupted everything, chilling a business that
was just beginning to heat up. In July 1914, with economic anxieties
increasing, Buffalo Forge decided to get out of the air-conditioning business and revert to its traditional manufacturing.[30]

"The entire country went into a tailspin," Willis Carrier recalled.

Carrier, however, decided to grab the controls. Along with his part-
ner, J. Irvine Lyle, Carrier boldly decided to start a new company; to hell
with the uncertainties.

With Willis Carrier as president, on June 26, 1915, Carrier Engineering Corporation officially was incorporated, Carrier, Lyle, and five
others assembling $32,600 in startup money. They set up outlets in New
York City, Boston, Chicago, and Philadelphia, with Carrier presiding in
two rented rooms in Buffalo and a collection of secondhand furniture

that included two wicker chairs. Friends asked if he had stolen them from a taproom. On July 18, Carrier signed its first contract, with the American Ammunition Co. of Paulsboro, New Jersey. The company's second job was the installation of a cooling system in a Masonic Temple in Philadelphia.[31]

The road to ubiquity had several turning points for the air-conditioning industry, and one of them was the cooling of the movie theaters. It has often been stated that Carrier began the movement with an installation in the Rivoli Theater in the heart of Times Square in Manhattan. And stated incorrectly.[32] The first "full air conditioning system" was installed in a theater in Chicago by Wittenmeier Machinery Co. in 1917. It was successful, and it was quite the draw.[33]

The Rivoli wasn't even Carrier's first movie customer. The company had cooled the Palace Theatre in Dallas and the Texan in Houston before taking on Times Square.

But Times Square was the blockbuster.[34] Carrier crowed that it was the Rivoli project that brought international attention to air-conditioned movie houses. The system had its first screening in 1925 on a warm Memorial Day, the traditional start of the summer season. Adolph Zukor, the Paramount Pictures head honcho, was among those in the audience. Due to some last-minute adjustments, however, the cooling machine was late in starting, and the theater was hot when the crowd filed in.

The engineers were sweating, and not just because of the heat. Zukor was paying at least as much attention to the patrons fanning themselves as he was to the picture. Eventually, the cool air routed the warmth, and a star was born or, more properly, gained more celebrity. When the movie was over, Zukor was heard to say of the mechanical marvel, "Yes, the people are going to like it."

They did.

Movie theaters became the precursors of the "cooling centers" that cities employ these days to provide relief from the heat. By the end of the 1920s, more than 300 movie houses were advertising that they were "cooled by refrigeration."[35]

Air-conditioning rapidly became an attraction in and of itself.[36] It was the agent of a revenue bonanza. It allowed theaters to stay open year-round, rather than just from November to May. It also gave the industry a signature institution—the "summer blockbuster." As a former chair of the Air Conditioning and Refrigeration Institute remarked, "People were trying to get away from the heat. . . . And they came in in droves."[37]

Air-conditioning also had a cosmic impact on the movie universe in another way, by aiding the development of talk movies. Filmmakers now could screen out outdoor noise by closing studio windows and doors when it was hot.[38]

Just as it was getting its legs, however, the air-conditioning industry had to negotiate yet another major obstacle. Having already contended with the Great War, Carrier and his colleagues were confronted with the Great Depression.

For obvious reasons, consumers weren't rushing out to buy air conditioners. Yet that evidently didn't stop the industry from making inroads in the 1930s. *Fortune* magazine published in an article in 1938 that presented a portrait of a growing phenomenon. It said that air-conditioning was "becoming a competitive necessity in summer wherever customers come to eat, drink, or buy. It has broken into office buildings, hotels, homes. It is almost taken for granted on Pullman cars. Ships, government buildings, hospitals have come to use it." The magazine praised Carrier and his associates for their salesmanship.[39]

British scholar S. F. Markham remarked in 1947, "The greatest contribution to civilization in this century may well be air conditioning—and America leads the way."[40]

The boom was just getting started. A giant step for the industry would be inside the doorways of the places where people lived. Residential units gained popularity during the prosperous years after World War II, with a million reported sold in just 1953.[41]

In 1960, air-conditioning made its debut in U.S. Census data. It was determined that 13 percent of all U.S. housing units had some form of air-conditioning.[42] The growth since then has been startling, buoyed further by the 1970s power surge of central air-conditioning.[43]

Today, nearly 90 percent of all residences are air-conditioned in some capacity, including 97 percent of them in the State of Delaware, and

96 percent in New Jersey, two states where summer humidities are prone to be oppressive.[44]

Air-conditioning would redefine U.S. demographics, making Sun Belt locations all the more appealing. The population of Phoenix more than quadrupled from 1950 to 1960, and 16-fold from 1950 to 2020. And check out Las Vegas. In 1950, it was home to 24,000. In 2020, the population was 632,000, a 16-fold increase.[45]

Air-conditioning unquestionably has changed America, but has it been for the better, and is the price of comfort dangerously exacting?

THE PRICE OF PROSPERITY

In the summer of 1979, as the air-conditioning movement mushroomed, essayist Frank Trippett argued in a *Time* magazine piece that it was extracting social costs. He wrote that it "seduced families into retreating into houses with closed doors and shut windows," resulting in the sunset of "the front-porch society whose open casual folkways were an appealing hallmark of a sweatier America."[46] I can't imagine anyone offering such a fascinating rumination these days, and not just because suburbanization and porchless homes built on sizable lots arguably have done far more to reduce the meetings of the "front-porch society." Trippett's comments preceded the EPA's landmark report on global warming by four years, and the brutal summer that would capture the attention of Congress five years later.

Concerns about air-conditioning have pivoted to its potential stress on power grids and contributions to greenhouse gases and, thus, its effects on climate as its proliferation and demand continue to increase worldwide. An analysis published in December 2021 concluded that to meet power demands in the United States, air-conditioning efficiency would have to increase by as much as 8 percent.[47] And all that power usage contributes further to the production of greenhouse gases, not to mention the consequences of the use of hydrofluorocarbon refrigerants.

Air-conditioning, thus, is "a double burden for climate change," observed Mark Radka, Chief of the Energy and Climate Branch of the United Nations Environment Program. "Right now, the more we cool, the more we heat the planet. If we are serious about reversing current trends, we cannot go about cooling our planet with a business-as-usual approach."[48]

And the worldwide air-conditioning movement shows no signs of cooling off. An additional 4 billion air-conditioning units could be added by 2050, with India and Indonesia accounting for a whole lot of them.

An essay in *Scientific American* posed the question: "Could the need for cooling wind up cooking the planet?" That can't be allowed to happen, but the world needs air-conditioning. Rather than "a threat," the writers argue, it should present an opportunity to explore greener and renewable energy sources.[49] Finding those options must happen right away, as a paper in the *International Journal of Biometeorology* concluded: "Due to the complex nature of the problem, there is no single solution to provide sustainable cooling."[50]

Having grown up in a house that didn't have it, sleeping in a bedroom where the only cooling came from a pathetic puff from a rotating fan that didn't do much to repel the mosquitos that invaded through holes in the window screens, I remain eternally grateful for air-conditioning. From my reporting, I know that heat is deadliest for the elderly, poor, and vulnerable who live in houses without it.

The rapid increases in the use of air-conditioning have been a source of consternation among some environmentalists, and understandably so. But in a study published in 2016, a group of researchers from Yale University, the University of Chicago, the University of California, Santa Barbara, Carnegie Mellon University, and Tulane University argued that air-conditioning is a critical resource in efforts to adapt to rising temperatures. They documented that since 1960, as many as 20,000 fewer heat deaths were occurring annually. "The diffusion of residential air conditioning after 1960," they concluded, "explains essentially the entire decline."[51]

AUTUMN
UNHEALTHY HARVESTS

Fall is what naturalist Edwin Way Teale called "the glorious sunset" of the year. For me, it is the most mystical of the seasons, when the oblique light of the sun transforms the colorful light of the woods into stained glass, when the setting sun electrifies the remnants of foliage on the treetops.

Every year I consult with my favorite veteran foliage experts to get a reading on how vibrant the colors will be. In real life, they happen to be serious scientists who have dispassionate reasoning behind their outlooks. In the end, however, they all agree that fall never disappoints, that at the end of the day, no matter how pallid we perceive the colors to be, the twilight and the trees will interact to work a magic that will match or outdo any daylight peak.

No one has figured out how to predict precisely just how intense foliage colors will be in any given area, in any given year, because so many factors contribute to inform nature's palette. They include the weather conditions of previous seasons; infestations; total rainfall or lack of it in recent periods; and the often-mysterious behavior of the stoic woodlands. These constitute a matrix of factors to rival the response of the human body to this season of profound and rapid change, and not just among the leaves.

Autumn begins by hosting the splendid residue of summer, when it's possible to enjoy seashore resorts without the summer crowds. Simultaneously, the ominous forces of winter are on the march. For millions of people, it is the season that is the most physically and mentally challenging, when our sense of well-being can parallel the weather for the rapidity of change. The retreating summer and the ambitious winter wage a titanic battle for supremacy in the temperate zones. Appropriately, it is fought most fiercely along "fronts" created when air masses advance from the colder regions of the planet, and meet their resistance from the tropical air fighting to defend its claims on the atmosphere. Where air masses collide, anything can and does happen as human beings are caught in the crossfire.

The ragweed-allergy season laps into fall, but the strong winds accompanying the passage of cold fronts actually can stir up other pollens on the ground, the remnants of the tree and grass seasons, that have since dried out and are available for air transport.[1]

Fronts can agitate and import a variety of particulate matter, with unpredictable respiratory consequences. I can attest to that phenomenon personally, and so can our older son, an asthmatic. When he was 12 years old, he had a life-threatening experience while on a school camping trip in Vermont near the end of September. He had a severe attack that so compromised his airways that he had to be rushed to an emergency room. The incident coincided with the passage of a potent front that stirred powerful winds.

As we will discuss chapter 9, for reasons that aren't entirely clear, the start of the school year is a harvest season for asthma, when children line up at school nurses' offices to seek relief with their inhalers.[2] Fall is a season for what is known as "thunderstorm asthma."[3] Fronts frequently set off convective thunderstorms whose cold downdrafts can be packed with particulate matter, including mold spores.

Allergists hold that ugly duckling mold spores are tremendously underrated as year-round tormentors, and autumn is their party season: They are apt to prosper on damp, fallen leaves and assorted foliage detritus.[4]

Frontal passages come with other health hazards. As we mentioned in the summer section, all thermal comfort is relative, and the sudden contrasts wrought by a rapidly moving air mass can cause blood vessels to constrict and, thus, raise blood pressure and be a danger for people with heart conditions.[5] One study found compelling evidence of links between fronts and aggravation of stomach ulcers.[6] But perhaps the source of the most powerful impact on human health is so obvious that in retrospect, it is hard to believe that it didn't receive serious attention until the late twentieth century.

Nature turns on the dimmer switch in autumn as sunlight becomes ever-more oblique and the light loss is rapid—more than two and half minutes of daylight per day in October in the temperate zones. The closing of the curtains affects us in ways conscious and unconscious.

I would argue that in so many ways, autumn is at once the most enchanting and the most humbling of the seasons, when we relearn that we do have some things in common with the trees and the birds and the chipmunks. Food cravings appear to increase, as does overeating: It's not just "the holidays."[7]

In short, we respond to light. Millions of us don't respond very well to the fading of the light, as psychiatrist Dr. Norman Rosenthal discovered. He had a name for this affliction: seasonal affective disorder.

CHAPTER 7

WHEN THE CURTAINS ARE CLOSING

- Seasonal changes in natural light can have tremendous impacts on our health and well-being.
- As many as 60 million Americans endure at least mild variants of seasonal disorders.
- Seasons have been changing for eons but, perhaps surprisingly, identifying and naming the syndromes is a relatively modern development.

+ + +

Save for the considerable Dutch influences on their histories, and populations that place them among the world's largest metropolises, Johannesburg, South Africa, and New York City are continents apart in infinite ways.

Having lived in New York and spoken with Dr. Norman E. Rosenthal many times over the years—his research being of professional and personal interest—I can attest that you would never confuse a New Yorker's accent with that of a South African.

But of all the differences between the two cities, one was of special significance for the Johannesburg native, and for the future of linking the environment to well-being. It was a matter of degrees—of latitude.

At a latitude comparable to that of Hollywood, Florida, and Baja, California, Johannesburg is about 1,000 miles closer to the Equator than the U.S. city that writer Washington Irving derisively called "Gotham," or "Goat's Town."[1] It is not exactly shocking that the cities would experience radically different weather. Snow and winter dampness, for example, are part of doing business in Gotham. In Johannesburg, where winter is a dry season, snow quite literally is big news.[2] Winter evenings primarily are sweater situations, more like April in Paris. "For all the turbulence of its politics," Rosenthal observed, South Africa "can truthfully boast about its climate."[3]

Not that Johannesburg was without seasons. Rosenthal knew summer was over when leaves began falling. Blossoms announced that it was sunset for winter. In spite of what he called the "mildness of the seasons, I was aware at some level of the effect they had on my mood." Rosenthal, who later would become a prolific writer, considered crafting a novel in which the mood of the protagonist changed with the seasons. While he eventually would put aside that idea, the concept foreshadowed the course of his career.

A DARK FUTURE

It was in the summer of 1976 that Rosenthal left Johannesburg to assume his residency at the New York State Psychiatric Institute. That was quite the mild summer by Gotham standards: Temperatures reached 90 only twice in August, peaking at an unexceptional 92.[4] That was a few shades toastier than the typical Johannesburg summer readings.[5]

However, what really vectored Rosenthal's attention was something that transcended weather, something that affected him like a natural drug. He marveled at the length of the summer days at the higher latitude, so far from the Equator. Closer to the Equator, the lengths of the days and nights don't vary that much through the year. The farther from the Equator, the longer the summer days, and the longer the winter nights. "The summer days felt endlessly long, and my energy was boundless."[6]

It was quite the contrast. The latest sunset of the year in Johannesburg occurs around 7:00 p.m., at the summer solstice, which occurs in December on that side of the world. About 14 hours and 40 minutes pass between sunrise and sunset.[7] In New York, the latest sunset occurs after 8:30, and daylight lasts 16 hours and 13 minutes. The sun doesn't set before 7:00 p.m. until September, right around the time of the equinox. "It seemed like just the place to be."[8]

But come fall in Gotham, Rosenthal discovered, the northern light sprang a leak. During September, the sunrise–sunset period shrank by more than 70 minutes, and by October, nights had overtaken the days in length by 35 minutes. Nights didn't last that long in Johannesburg until the approach of the winter solstice.[9]

The effects on the South African transplant were dramatic.

Even the New York sunlight was different in autumn. "When the sun shone, its rays struck the earth at a strange oblique angle," he said.

On November 1, 1976, Rosenthal had what to him was a disorient-ing, shocking experience. An alarm went off. The clocks had returned to standard time the day before, and by the time he left work that Monday, it was dark. Foreshadowing what was to be a profoundly cold winter,[10] a stinging northeast wind added bite to the premature chill as temperatures fell to the upper 30s not long after sunset.[11] It was the time when the fallen, brittle leaves skitter across the ground and the hard-topped streets and sidewalks of Manhattan. He wondered how he could have been so energized during the summer.[12] "Suddenly, I had a sluggishness and a lethargy and difficulty creating and producing that I had not experienced before."[13]

All things must end, and that does include the bad. When Rosenthal experienced his first New York spring, his energy surged again.[14]

He knew he was on to something.

A "CRITICAL STEP"

As so often happens in the pursuit of a mystery, way leads upon way, and the results can rely as much on serendipity as determination. That was the case with Rosenthal. At a party in New York he had a chance meeting with psychiatrist Alfred Lewy, who informed him that he and colleague Tom Wehr had just made an important discovery about the human hor-mone melatonin, which the body produces in darkness.[15]

Melatonin is still guarding a treasury of secrets, but efforts to unlock them would be a "critical step" in the findings regarding how humans respond to seasonal changes and in developing treatments.

Rosenthal would leave New York to accept a research fellowship at the National Institute of Mental Health (NIMH), in Bethesda, Maryland, where he joined both Lewy and Wehr. They pursued the questions that piqued their mutual interests. Yet, according to Rosenthal, it wasn't a high-powered psychiatrist who would make the most significant contribution to the research that led to identifying seasonal disorders. It was an engineer.

His name was Herb Kern, a wiry, youthful-looking man with no medical training. He had read about the institute's explorations and came to it seeking help with emotionally crippling conditions that beset him every year. Kern had kept scrupulous records of how, with the annual waning of the light, without fail the energy leaked from his body. He had trouble making decisions, lost interest in sex, and wanted to "withdraw from the world." He had been treated with antidepressants, only to experience horrific side effects. The only true relief he found was in the lengthening of the days, when "the wheels of my mind began to spin again."

It happened that a second patient, identified by the pseudonym "Bridget," who described herself as "a human bear," appeared at the institute with quite remarkably similar complaints.[16] With a second case in hand, Rosenthal was confident that he and his team were getting somewhere, but in the world of research, twice is not enough.

He needed to cast a net. He decided to approach a journalist, engaging in what he called a form of "ambulance-chasing."[17]

Ouch! Norman.

THE BURSTING NET

That journalist was the late, and influential, Sandy Rovner, a *Washington Post* health columnist.

The article Rovner produced, which was published on June 12, 1981, began with a quote from Bridget: "I should have been a bear." (I should mention that in my wire-service days, we were told we could use one quote lead a year. I gladly would have used this one.)

The article contained extensive observations from Rosenthal and Dr. Frederick Goodwin, under whom he studied, and discussed the institute's research at length. The "bear" subject in question, "says she was like the grasshopper who played through the summer, forgetting that the winter had ever been, until the fall came," Rosenthal related.

Rovner ended the piece with comments and an open-ended plea from Rosenthal. In the newspaper business, this appeal would have been fair game in a column.

Rosenthal announced that the NIMH team was seeking people "who have gone through enough cycles to know that this is going to happen again." The researchers, he said, "would like to hear from anyone with distinctly seasonal mood disorders. Applicants will be sent question-naires, from which participants will be selected."[18]

I suspect that this piece would have generated far more responses had it been published in November or December. That notwithstanding, what followed was a dam burst. Thousands upon thousands requested the surveys. More astonishing than the volume, recalled Rosenthal, was the uniformity of the answers.

"I slow down when October comes. . . . I sleep too much. . . . I eat more than usual, especially sweets and starches. . . . I gain weight. . . . I can't concentrate. . . . I become depressed because I fail at various things. . . . I withdraw from friends and family. . . . Again and again and again I heard these things, and I thought: Wow, we've got a syndrome over here. It would be too much coincidence to be answering these questions in the same way."[19]

"As I read the questionnaires, it seemed as though Bridget had been cloned."[20]

THE NAMING

How the seasons affect moods was hardly a new concept, but it was Rosenthal who gave the extreme form of this phenomenon a name—seasonal affective disorder, with the almost too-perfect acronym, SAD.

The term first appeared in a 1984 paper by Rosenthal and his NIMH colleagues.[21]

SAD is a form of depression, prevalent mostly in the light-deprived cool seasons, although some people have been diagnosed with the summer variant.[22]

The milder form of seasonal affective disorder is called "winter blues." In fact, Rosenthal's popular treatise on the subject bears the title *Winter Blues*; however, the onset of the symptoms typically occurs in fall as the light rapidly dwindles.

Numbers vary on how many people are affected by SAD, but Rosenthal has estimated that as many as 13 million Americans suffer from SAD-related depression, for which symptoms can persist as long as five months. Up to 30 million may be afflicted with the tamer version.[23]

Given its annual occurrence and duration, SAD "is considered a serious mental health problem" in the psychiatric community.[24]

Underscoring that the loss of light is a prime factor, SAD is seven times more prevalent in Washington State than in Florida. Women patients outnumber men, four to one. Up to 20 percent of the population suffers from a mild variant of SAD.[25]

Clinical diagnoses aside, the fading of the light and the onset of colder weather in the populous midlatitudes indisputably have profound mental and physical effects, despite the technological advances that have insulated us from the harshest impacts of the elements.

"Light is a very powerful drug," says Dr. Phyllis C. Zee, a professor of neurology and chief of the Center for Circadian and Sleep Medicine at Northwestern Medicine, in often light-deprived Chicago. "We don't think of ourselves as being seasonal creatures, but we are. There are changes that occur in our metabolism. There are changes that occur in our immune system over the course of the seasons." Although they are not as pronounced, seasonal effects are evident even in places "that you would not consider so seasonal," such as parts of California.[26]

As Rosenthal pointed out, "to a greater or lesser degree, a great majority of the population experiences some changes in seasonal well-being and behavior." This might involve changes in energy, sleep patterns, mood, or eating habits.[27] Yes, even eating habits.

As so much in biometeorology, quantifying and teasing out the contributions of the natural environment to "well-being and behavior" is an immensely complicated exercise. But one fascinating study published in 1991 provided a morsel of one of the classic symptoms of the human response to waning light.

IT'S NOT THE HOLIDAYS

The "holidays," which begin more or less at Thanksgiving and continue into the new year, have often been linked to excessive eating and drinking and, thus, weight gains. In an era when obesity is a borderline epidemic, that's a rather serious indictment against what should be a festive period. But in one of the grandest experiments on record, psychologist John M. de Castro, whose specialties include American eating habits, made a compelling argument that the holiday pork-up is a myth.

It is really about the seasons.

In de Castro's study, 315 adults were paid to keep seven-day diaries of everything they ate and when, and to note just how hungry they were. The holiday period was excluded. What he found was "a marked seasonal rhythm of nutrient intake, especially of carbohydrates, in the fall." Of the four seasons, the subjects consumed substantially more calories in the fall than they did in the other three. Compared with the other seasons, they ate larger meals. They ate faster, and they reported they were hungrier afterward even though they had eaten more. "The results suggest that even with modern heating and lighting, seasonal rhythmicity of food intake persists in humans and is a major influence on eating."[28]

People eat more in the fall as the solar energy reaching the earth wanes.

BEAR FACTS?

Rosenthal, who speaks with a psychiatrist's calming voice, mellowed all the more by a Johannesburg-cultivated accent that has survived half a century in New York and the Capital Beltway, has given the question intensive thought: Are seasonal effects all about the so-called hibernation response? It's a nature question, and "hibernation response" understandably shares with "winter blues" a popular label for SAD and its variants.

This drive to be "recumbent" and "to be left alone in some dark space" evokes the image of a bear in a cave, as Rosenthal observed in *Winter Blues*. But, clearly, humans and bears have little in common regarding how they pass their winters.

Contrary to the popular conception, bears don't hibernate in the true sense, those who know the ways of bears inform us. Bears don't spend their winters sleeping but, rather, enter a state of "torpor," a sleep-like state in which their heart and breathing rates slow dramatically. When they do sleep, it is profoundly deep. During the winter, when food and water are scarce, this is how bears conserve energy.[29]

The humans I have known aren't very bear-like. They keep right on eating once it turns colder, and for most of them food scarcity does not appear to be an issue, based on what I've seen in advertisements and in the proliferation of fast-food restaurants and health-conscious markets. "We don't have such a thing as, we don't have food in winter," says Dr. Zee.[30] Nor do humans have much in common with true hibernating mammals, such as chipmunks, whose heart rates drop from 350 beats per minute to a rather unbelievable four beats per minute.

But those obvious dissimilarities don't negate the legitimacy of the question of whether SAD may be some form of energy-conserving "adaptive" response, says Rosenthal.

Is society's expectation that we maintain the same levels of activity and energy year-round unreasonable? "We have no good answers to these questions," Rosenthal concluded.[31]

THE "SUPRACHIASMATIC" EXPLANATION

What makes us behave differently with the procession of the seasons? Like other mammals and the trees and other forms of plant life, humans do have adaptive responses to the diminution of natural light, almost as though energy levels were dependent on so many solar panels.

Dr. Zee has a rather grandiloquent explanation. "Our body," she says, "including all of our organs, is aligned to match the environment. It's a beautiful symphony orchestra." The conductor, to complete the metaphor, is the brain's "suprachiasmatic nucleus." That is quite a mindful.[32]

In lay terms, that suprachiasmatic nucleus—or the SCN—is the "central pacemaker" that regulates most of the body's circadian rhythms.[33] It relays messages to other parts of the brain about the time of day, picking up on light cues. It is composed of about 20,000 neurons.[34] (While that may seem like a whole lot of neurons, on average a human being's brain has 86 billion of them, give or take a few billion.)[35]

LIGHT AND HORMONES

How all this messaging and circuitry operates is a marvel, and for what we know about it all, we can thank legions of neurological researchers. They tell us that with decreasing natural light, our bodies are under orders to shift to a lower gear, and the increasing darkness triggers significant changes in three important hormones. It reduces levels of mood-boosting dopamine and serotonin, and stimulates increases in melatonin, which can make us tired at inopportune times.[36]

"DOPE" AND DOPAMINE

"Why do you think they call it dope?" That was a wildly popular anti-drug-abuse slogan promulgated by the Advertising Council from 1965 to 1980 when illicit recreational drug use was rampant across the country.[37] (That campaign included marijuana, obviously predating the era of decriminalization and widespread legalization.)

According to a variety of sources, "dope" is a derivative of a Dutch word for "sauce." It would be perfectly natural to believe, however, that "dope" was slang for dopamine, the "feel-good hormone."[38] Dopamine is a key part of our evolutionary reward system. When we are engaging in pleasurable activities, say, eating something we enjoy or engaging in sex, our brains release dopamine, and we want more of that stuff.[39] Dopamine deprivation can lead to mood swings and, in some instances, even depression.[40]

"RUNNERS' HIGH"

I never got to the marathon level, and I have no regrets, but for several years I was a 10-mile-and-change runner. I deeply enjoyed it, and like so many things in life, I especially liked it when I stopped and savored the euphoria.

I was put off one day when an older neighbor all but accused me of being a drug addict. He suggested that all runners wanted in life was the hormones that were shaken loose by the brain due to all that knee-pounding.

Personally, that's an addiction I would welcome anytime. What goes on inside the bodies of runners and others involved in intense physical activity, such as bicycling and weightlifting, is immensely complex and not entirely understood. You'll find different informed opinions and analyses, but serotonin is very much a factor, explains Stephanie Watson, former top editor of Harvard Women's Health Watch. During workouts, the body releases quantities of tryptophan, the raw material the brain uses to make serotonin.[41]

How important is serotonin to well-being?

The mission of some popular antidepressants is to increase serotonin levels in the body.[42]

DARKNESS, DARKNESS

While the waning of the light and increasing sunset-to-sunrise periods deplete dopamine and serotonin, they have the opposite effect on melatonin.

The darker the room, the higher the melatonin production. Like so much pertaining to hormonal processes, melatonin's multiple roles in health haven't been fully identified. However, it is called the "sleep hormone" for its role in regulating those 24-hour circadian rhythms so instrumental in determining when we eat and sleep, and when we are most alert.

So many factors play into our sleeping habits, but it is known that we sleep better when our bodies have the peak levels of melatonin, which is all about darkness.[43] Melatonin, produced during the dim light of those darker days, lets the air out of our energy and alertness balloons.[44] Nap, anyone? Melatonin is a pivotal player in our well-being—and is at the heart of the pursuit of remedies for seasonal affective disorder.

For Rosenthal, it started with that chance meeting in New York. Alfred Lewy had attended that aforementioned party in New York because he was dating someone in Rosenthal's program at Columbia.[45] Lewy had developed a method to monitor melatonin levels, which occur in minute quantities in the body, by using a spectrometer that looked not unlike "a very large washing machine."[46] At the party, he told Rosenthal that he and Tom Wehr had discovered they could suppress melatonin by exposing subjects to bright light.

It was known that light drove seasonal rhythms in other mammals. Could it also be driving human seasonal rhythms? Herb Kern, the engineer, helped to answer that question emphatically. His careful chronicling of his symptoms through the year was compelling and convincing. Clearly, whatever was happening to him was the result in the changing of the light.

If that were the case, would exposure to light relieve his symptoms? The institute team and Kern gave it a shot. During the darker days, the team exposed Kern to artificial light for periods that were equal to summer day lengths. "Sure enough," said Rosenthal, "he came out of his depression."[47]

I haven't spent much time in my life meditating on the physiology of how all this works, but as a card-carrying atmo-chondriac, I can say that when nature turns down the dimmer switch, I get the message. I am grateful for the neurologists and psychiatrists who have explored the intricacies and produced a vast volume of literature and academic treatises on the subject, including Dr. Rosenthal. His *Winter Blues* is full of insights and serves as a veritable instruction manual on how to recognize and combat negative reactions to the seasonal erosion of natural light. I also have admired his taste in literature—check out his Poetry Rx— which is evident in *Winter Blues,* in which he included two wonderfully descriptive passages that capture the essence of the darkest moods that can overcome one as autumn closes the curtains.

One was from a letter written by Henry Adams: "It is one of the darkest, foggiest, and dismalest of November nights, and as usual, when the sun doesn't shine, I am as out of sorts as a man may haply be, and yet live through it." The other was a rumination on Ishmael in Melville's

Moby-Dick: "Whenever it is a damp, drizzly November in my soul; whenever I find myself pausing before coffin warehouses, . . . I account it is high time to get to sea as soon as I can." (In Ishmael's case, given the outcome, I suspect Dr. Rosenthal would have recommended that he choose light therapy over a sea venture.)[48]

While it may be more normal to believe that we transcend our natural environments than it is to acknowledge that these affect our well-being, Rosenthal is among those who emphatically believe that our beings are tethered to the magical progression of the seasons. As he wrote in *Winter Blues*, "Like the birds, bears and squirrels, humans have evolved under the sun. We incorporated into the machinery of our bodies the rhythms of night and day, of darkness and light, of cold and warmth, of scarcity and plenty. Over hundreds of thousands of years, the architecture of our bodies has been shaped by the seasons, and we have developed mechanisms to deal with the regular changes they bring. Sometimes, however, the mechanisms break down."[49]

If you do experience troublesome symptoms as the days get shorter, here are recommendations from Dr. Zee and experts at the University of Michigan in Ann Arbor, where sun is very much the exception November through February.

+ + +

- **Get out there.** Get into bright light as soon as you can after waking up, says Dr. Zee. Morning light appears in an optimal spectrum and "is more alerting."[50] Getting out of bed around the same time every day is ideal for keeping your circadian clock in sync. Seek natural light exposure throughout the day.

- **Get moving.** Exercise is critically important in supporting mood, especially in the fall and winter, when we tend to be more sedentary. Consider going for a walk or doing some type of workout indoors, whether it's aerobics, yoga, or hopping on a treadmill.

- **Get adapted.** When the light fades, the temperatures tend to drop, but we can adapt. The sooner you adjust to the cold by getting outside, even in the fall, the easier it is to get out later in the winter. To quote the Michigan folks, "There's no such thing as bad weather. There's only bad clothing."[51]

- **Get a light box.** If need be, opt for bright-light therapy, which you can get through using a light box or light glasses for 15–30 minutes in the morning.

- **Get help.** If you're having trouble engaging in your usual social activities or focusing at work, those could be warning lights for SAD. Talk to your mental health professional if this is the case. You don't have to struggle alone.[52]

CHAPTER 8

FANFARE FOR THE COMMON COLD

- In all likelihood, colds have bedeviled humans since prehistoric times. The first written account appears in Egyptian writings about 3,500 years ago.
- Exposure to cold weather was long believed to be the cause of the common cold. Benjamin Franklin was among those who thought that idea was bunk. Yet even today some people aren't convinced.
- Working with 20,000 volunteers, a research group spent over four decades trying to identify the viruses that cause colds. They did not find a cure. Nor has anyone else, but chicken soup can help, for real.

+ + +

After all these millennia of human existence, it's still true that everyone gets "colds," but no one yet has figured out how to cure them. They are very much weather-related, but it took thousands of years of misconception to figure out how.

You're likely to see the words "mild" and "harmless" associated with this affliction, but its direct and indirect economic costs are staggering, an estimated $80 billion annually in the United States.[1] They also can be gateways to sinus and ear infections and acute bronchitis. For people with respiratory conditions, such as emphysema, they can be particularly burdensome.[2]

And on a personal level, I have long held that the cold remains one of the most underrated ailments in the annals of medicine. I speak from experience; I've had more than my share, especially when our kids were in the K–12 universe when they were always diligent about sharing whatever was going around in the schools.

The disrespect for colds is boundless. If you have one, people expect you to show up for work and act as though you don't have one—even if they very much act like you do, and give you the leper treatment. "The common cold is so common that we are apt to pass it by with a contemptuous gesture, unless, of course, we are the sufferers," to quote a 1931 editorial in the *Canadian Medical Association Journal*.[3]

Colds are akin to snowflakes in that they can be very much alike, but no two are identical, and with good reason. Yet they all have a certain irritating rhythm, from onset, to outward symptoms, to resignation, to that unacknowledged debilitation.

It's never like it is in the movies with some actor or actress doing a horribly phony rendering of sneezing or coughing and trying to affect a "cold" voice. In real life, colds begin typically with a vague and discomforting sensation that clouds are gathering in the head. That is followed by what I call the "dirty windows" syndrome, in which the visual powers of the mind are obscured. What follows is the paradox of unrelenting stuffiness and nonstop running of the nose that can transform the simplest chores into projects. I invariably lose my ability to smell—not always a great loss—and taste—which I very much mind. All alcohol, for example, tends to taste like Listerine.

And the worst of it is, for all this enduring and suffering, no one gives a damn.

That includes my wife, and in her case, that's excusable. She has made it clear that she has no interest in my hourly symptomatic updates. When she is getting a cold, she doesn't complain, or even talk about it. When I ask how she is feeling, she will accuse me of being concerned—not about her health, but about my prospects of getting what she has. OK, so she caught me.

Her expressions of sympathy are limited to: "It's a cold. It's not cancer."

COLD FACTS

In the universe of diseases and ailments, a signature feature of the "common cold" is the adjective. Ron Eccles, who for 30 years ran the Common Cold Centre at Cardiff University in the United Kingdom, says that to an extent the adjective is well-earned. "It's part of our culture," he says. "It's part of the folklore because everyone has views on the common cold." It is one illness that is truly intergenerational. Even children can tell you all about it, "since they've had so many."[4] But Eccles argues that the cold is "a cultural concept rather than a clinical entity as the disease is usually self-diagnosed and treated by the patient."[5]

Pared to its basics, a cold is an upper-respiratory-tract illness, affecting the throat and nose. Yet it is caused by as many as 200 different viruses.[6] "I don't often call it a specific disease," said Eccles. "It's more like a syndrome of symptoms that we call the common colds."[7] Those symptoms usually ripen one to three days after exposure to the attacking virus, and persist for up to 10 days. Along with that runny-nose/stuffy-nose paradox and loss of taste, familiar symptoms include a sore throat, coughing, and perhaps a low-grade fever.[8]

That modest list of petty torments empties into another cosmic question: Why on earth do we call it a "cold"?

According to the Online Etymology Dictionary, the term dates to the sixteenth century and was derived from the fact that "the symptoms resemble those of exposure to cold."[9]

Really? Wasn't that during the Renaissance? I thought people were getting smarter about then.

When I've had colds, I didn't necessarily want to take walks in the snow, or take walks, period. But I can't say that the discomforts attendant to the illness paralleled those of being in the cold for long periods, an experience that under the right circumstances I have even enjoyed.

Evidently, that misconception about colds has had real staying power. To this day a whole lot of people continue to believe that cold weather actually is a cause of the common cold, as witnessed by any number of wrongheaded literary and media references.

Affirming a truth that Ben Franklin held to be self-evident, the experts at Johns Hopkins Medicine state categorically that "contrary to popular belief, cold weather or being chilled doesn't cause a cold." It is true that the cooler temperatures increase the potential for viruses to spread. However, that's because the air becomes considerably drier than it is in the summer, making it more favorable for airborne viruses.[10]

I'm with Franklin and Johns Hopkins. If I bought into the association of the common cold with the stuff that makes us shiver outdoors, this chapter would be in the winter section, where you'll find the discussion of influenza.

A word about influenza: I've had three episodes of flu in my adult life, all in winter. One big difference between cold and flu for me was that when I had the flu, I had neither the energy nor interest in discussing my symptoms. I believe I'm on safe ground in deciding that the severities and durations of flu symptoms so far surpass those of colds that they belong in separate chapters. Tragically, flu has been implicated in some of the deadliest disease outbreaks in world history. No one ever accused the common cold of being a mass killer.

Besides, the flu season more properly belongs in winter because it typically doesn't get going until December and doesn't peak until February.[11] By then, chances are that you've had a cold or two, in which case you may have a shared experience with any number of Egyptians of ancient times.

SNEEZING ON THE NILE

In all likelihood, humans were catching colds before anyone bothered to write down or record anything.[12] As best as can be determined, the first written account of a cold likely dates to about 4,000 years ago.

The Egyptians didn't have a *Physicians' Desk Reference*, but it is believed that they produced the world's first medical tomes. They were discovered by George Ebers and Edwin Smith in the nineteenth century. The so-called Ebers papyrus is all of 200 feet long and contains write-ups of "numerous illnesses" and treatment recommendations. It also describes 20 different types of cough, complete with a cold-breaking spell.[13]

Hippocrates, who probably holds the record for medical "firsts," appears to have been the first to describe cold symptoms. "When we have a running of the nose and there is a discharge from the nostrils the mucus is more acrid than when we are well," he wrote. "It makes the nose swell, and renders it hot and inflamed."[14] Hippocrates predates me by a couple thousand years, but I couldn't have described these symptoms any better myself.

The Roman doctor Celsus, who lived in the first century CE, discussed symptoms that unquestionably were brought on by a cold. "This closes up the nostrils, renders the voice hoarse, excites a dry cough."[15]

Hundreds of years later, Arabic physicians remarked on contagion and colds. "Contagion is a spark that flies from a sick body to a healthy one," observed Ousta the Luga, who is believed to have lived from 830 to 920 CE. At the time it was thought that the winter chill was the agent behind the common cold. Sound familiar? However, Moses Maimonides, who lived about 200 years after Ousta the Luga, addressed the phenomenon of the summer cold. Obviously, that was unrelated to cold air, but rather to the heat "that melted the hard excretions that are found in the brain, and then they run down."[16] That suggests that summer-cold sufferers could blow their brains out in a benign way.

Deeper than the question of who first called a sneeze a "sneeze" is the mystery of the identity of the Renaissance genius or geniuses who first started calling these congestive annoyances "colds." But a few centuries later, the observations of one of the most inventive figures in American history represented a huge step in disassociating the common cold from the cold outside. and opened a window of understanding.

REVOLUTIONARY IDEAS

Benjamin Franklin famously documented the connection between lightning and electricity,[17] but Franklin rivals Hippocrates for the numbers of "firsts" on his resume. The inventions credited to him include the lightning rod (of course), the Franklin stove, bifocals, and swimming fins. He founded or cofounded America's first public hospital, first circulating library, first volunteer fire company, and first mutual insurance firm.

Being a weather geek, I am especially impressed that he figured out that storms moved in directions counter to the winds blowing at the surface, and that he managed to chart that all-important weather-maker, the Gulf Stream.[18]

All that accomplished, I was to learn that he was among the first to separate cold weather from the common cold. The source of this revelation was John Adams, who recounted it in a journal entry in which he expressed skepticism about Franklin's hypothesis. In fact, Adams seems to suggest that his friend and fellow Founding Father was a bit of a windbag. The entry, interestingly, was dated September 9, 1976.

You might think that those two would have had other things on their minds at the time, say, *Common Sense* rather than the common cold. They had, actually. Two months after boldly declaring independence, and 17 months after the first shots were fired in the Revolutionary War, Franklin and Adams were part of a delegation that was to meet with British General William Howe to discuss the possibility of a truce.[19]

On the journey, Adams was appalled that "on the road and at all the public houses, we saw such numbers of officers and soldiers, straggling and loitering. . . . Such thoughtless dissipation at a time so critical, was not calculated to inspire very sanguine hopes. . . . I was nevertheless determined that it should not dishearten me."

Along the way the pair stopped in New Brunswick in northern New Jersey. They spent the night at an inn. It wasn't a Marriott. The lodgings were so tight that he and Franklin had to share a bed "in a chamber little larger than the bed, without a chimney and with only one small window." Being "afraid of the air in the night," Adams shut it. "'Oh!' says Franklin, 'don't shut the window. . . . Open the window and come to bed. . . . I believe you are not acquainted with my theory of colds.'"[20]

In the popular images Franklin typically is depicted as a rather rotund, aging figure who never had been a gym rat. Yet he was strong and fit in his younger days. If he possessed an Achilles' heel, it was his struggles with respiratory conditions, which had fired his interests in colds. He was convinced that the popular notion that they were caused by exposure to cold air was pure bunk.[21]

That night in Jersey, Adams allowed that he was well aware of the Franklin hypothesis, since he had read Franklin's letters on the subject in which the famous Philadelphian argued that "nobody ever got cold by going into a cold church, or any other cold air." Adams, however, was not convinced. "The theory was so little consistent with my experience, that I thought it a paradox." Nevertheless, he allowed Franklin to proceed with his lecture.

Franklin held that by closing the window and cutting off the outside air, the two essentially would be breathing recycled air and would "imbibe the real cause of colds."

Adams endured the argument, and concluded, "There is much truth, I believe, in some things he advanced, but they warrant not the assertion that a cold is never taken from cold air. . . . I believe, with him, that colds are often taken in foul air, in close rooms. But they are often taken from cold air." Nevertheless, Adams said, "I had so much curiosity to hear his reasons that I would run the risque of a cold." Franklin "then began a harangue, upon air and cold and Respiration and Perspiration, with which I was so much amused that I soon fell asleep, and left him and his Philosophy together."[22]

That they couldn't reach an agreement after an amicable discussion foreshadowed what was to come on September 11 on Staten Island when they met with General Howe. That, too, was an amicable discussion, but they could not reach an agreement with Howe, nor could they reach an agreement with the British generally on an armistice.[23]

The war would drag on.

During the famous encampment at Valley Forge, illnesses claimed the lives of 3,000 soldiers. The privations of the winter of 1777–1778 were legendary. As for any role of colds, almost 70 percent of those Valley Forge fatalities occurred during the warm seasons.[24]

THE COLD COMES IN FROM THE COLD

The journey to identify the real causes of common colds was laborious and labyrinthian. They once were attributed to evil spirits and "body humors."[25] In an 1826 article in *The Lancet*, a Mr. Lawrence of the London Ophthalmic Infirmary offered a tidy summary: "The chain of causes seems equally unknown to the learned and in the ignorant."

But eventually the word "cold" would achieve status as a homonym to denote both an illness and something for which one might require a coat.

It was Louis Pasteur in the mid-nineteenth century who developed the germ theory of disease transmission.[26] And in Germany in 1876—a hundred years after Franklin had lectured John Adams on the causes of colds—Robert Koch became the first to prove that a micro-organism could cause a disease.[27] The common cold at long last was coming in from the cold. Yet old ideas about the source of the infections would survive with the tenacity of a head cold.

THE COLD TRUTH

It took until the mid-twentieth century to establish that certain respiratory viruses were the agents that set off the discomforts attendant to the common cold.[28] One reason for the delay simply was the fact that researchers had other fish to fry. The common cold, as the adjective suggests, was not in a league with the likes of life-threatening—and life-taking—typhus and influenza.[29] This is perhaps understandable; lack of respect was a long-standing tradition. As that 1931 Canadian journal editorial suggested: "That it has not received more attention is probably due to the fact that the affection is transient, is borne with as a trifling and passing evil, and no deaths are the result of it."[30]

Eventually, however, the cold would have its day.

Under the British Medical Research Council, a Common Cold Unit was established under Sir Christopher Andrewes and Dr. Alick Isaacs.[31] If you're wondering why the United Kingdom became such a hotbed of cold research, it so happens that the English apparently have had a considerable history of nasal congestion. "The French used to have a name for the Englishman," said Eccles. "It translates to 'a Man with a Stuffy Nose.' I think we do get more colds than the French."[32]

In Andrewes's words, the group aimed to take "another crack at the nut" to get at the kernel of what was causing a prosaic illness that was bedeviling the English and seemingly affected every human being multiple times per lifetime.[33] The unit set up shop in 1946 at Harvard Hospital, in a countrified setting near Salisbury, outside London. That name may sound familiar, and indeed it was a gift of Harvard University and the American Red Cross in 1940. Its original mission was to serve as a 125-bed hospital to control anticipated war-related epidemics. Five years later, the hospital and other buildings on the campus were donated to the British for communicable-disease research, and the council took over the buildings for the common cold project.

"YOU MAY NOT WIN A NOBEL PRIZE"

While the search for a cure would end up being a cold trail, it would yield a treasury of knowledge about colds. The quest would involve 20,000 volunteers who risked their noses and throats for the sake of one of the grander medical investigations in medical history.[34]

Andrewes recalled that the researchers "groped in the darkness of their ignorance. . . . In our artless 1946 way we talked about 'the common cold virus,' fully aware that there might be several of them."

They had chased fantastic reports from people who "told us of animals which caught colds—cats, vervet monkeys, capuchin or weeper monkeys, hedgehogs and flying squirrels." The reports were given serious treatment. "All these were tested, but without producing any evidence that their colds were like our colds, or that our viruses would do anything to them."[35]

Given the times, investing money and resources and brain power into an investigation of the common cold was an unlikely project, observed Dr. David Tyrrell in his wonderful book, *Cold Wars*, cowritten with Michael Fielder. Tyrrell shared incredible detail regarding goings-on at the Harvard campus, and he was a most credible witness: He would lead the unit for over 30 years.[36]

He noted that like so much of the world, the British were emerging from war and were confronted with an unprecedented rebuilding project. Under the circumstances, he noted, a project of this nature wouldn't have been the empire's number-one priority, and, again, this was not an affliction in the category of flu and typhus. However, the war "had shown how science could solve problems and introduce new ideas." For examples, he cited radar and atomic energy. "It was the right climate in which to consider how medical science could be mobilized to tackle the problems of disease."[37]

After considerable discussion and preparation, project cold was ready for launch. "Everyone was fascinated to see how this historic investigation would proceed," Tyrrell said.[38]

An important first step was the casting of the net for volunteers. One advertisement for volunteers promised a "Free 10 Day Autumn or Winter Break: You May Not Win a Nobel Prize, But You Could Help Find a Cure for the Common Cold."[39]

The first volunteers arrived on July 11, 1946. The subjects, who were reimbursed for travel expenses, were quarantined for three days to make sure they didn't bring any colds with them. A fluid, produced in a lab and "developed" from a cold that affected one of the scientists, was "dropped into the noses" of some of them. Others received a dose of something harmless. The volunteers weren't told which dropping devices were which. "Watching the symptoms over the next five days was the key test."[40] The experiments were successful in that they established that colds could be induced by introducing a cold virus.

One may reasonably wonder whether this was an exercise in sadism and the subjects were somehow masochistic. According to Tyrrell's account, some very much enjoyed their respites in the country. One woman came back six times in 11 years. Another declared, "I was able to retreat in the monastic sense." And, according to Tyrrell, "Surprisingly, many of the volunteers were quite eager to catch a cold, even to a point of imagining the symptoms."[41]

Perhaps more surprisingly, most of the volunteers were unsuccessful. Those who did catch colds blew their noses into paper handkerchiefs that then were weighed and counted.

THE RESULTS

By the time the project shut down in 1990, researchers had learned that it wasn't a solitary virus that caused colds, but 200 of them.[42]

While assorted rhinoviruses likely were the cause of most colds, Andrewes concluded that "the rhinoviruses turn out to constitute a family of many antigenic types." He noted correctly back in 1966 that new viruses were "being added at a great rate," and it appeared that different variants appeared in different years. For all the variants and ubiquity of colds, Andrewes took note of the paradox that while colds are everywhere, they are "poorly infectious, in contrast to influenza, measles and other viruses." He surmised the viruses were subject to a Goldilocks factor: They had to be just right, that is, small enough to travel through the air, big enough to survive the flight, and not so large that they fall to the ground before finding a host. He theorized that viruses were more likely to cause colds in people who were subjected to "some local stress, weather change or other circumstance, in a state of temporary vulnerability."[43]

From my experience, I would say he was on to something.

WHERE IS THE CURE?

Although it wasn't exactly a tribute to the medical profession, Tyrrell was particularly fond of a poetical treatment by humorist A. P. Herbert.[44]

To wit:

I love the doctors—they are dears; But must they spend such years and years Investigating such a lot of illnesses which no one's got,
when everybody, young and old, Is frantic with the common cold?
And I will eat my only hat If they know anything of that![45]

If he were true to his threat, his hat would have become part of his digestive system: In fairness, doctors have learned a whole lot about colds.

It was Tyrrell who identified most of the cold-causing viruses. But one thing the cold unit's researchers learned was that it would be all but impossible to develop a common-cold vaccine that would be effective. The reason, very simply: The viruses come in so many varieties. As Andrewes wrote over a half-century ago, "what is quite clear is that even though a vaccine may be effective in protecting against" one variant, so many are out there that creating a vaccine would be a "formidable proposition."[46]

Sorry, Mr. Herbert, it still is.

"The problem was they were too successful," Ron Eccles said of Andrewes and Tyrrell and the unit's researchers. "They didn't find one virus. They started to find many viruses. This is not a simple disease like chicken pox or measles, or small pox, or polio. . . . It is one of the most complex diseases, despite being the most common."[47]

+ + +

Medical experts suggest that you never give a cold virus an even break. Wash your hands early and often, be chary about touching the face. The hands are a favorite commuting option for cold germs. Do those things you should be doing anyway, as in exercising, eating wisely, getting plenty of sleep, and relaxing when opportunities allow, all to build up your immunity.[48]

If you do catch a cold, as is prone to happen, here's advice from the Mayo Clinic to go along with over-the-counter meds that can relieve symptoms:

- **Drink.** Non-alcoholic liquids, that is. Water, juice, clear broth, or warm lemon water with honey can help loosen congestion and prevent dehydration. Avoid alcohol, coffee, and caffeinated sodas, which can make dehydration worse.

- **Rest.** No better medicine out there.

- **Throat relief.** To tame a scratchy throat, try gargling 1/4 to 1/2 teaspoon salt dissolved in an 8-ounce glass of warm water. Honey and hot tea can help tame a cough.

- **Nasal unclogging.** Over-the-counter saline nasal drops and sprays can help relieve stuffiness and congestion. Add moisture to a room with a cool-mist vaporizer or humidifier.[49]

CHAPTER 9

ASTHMA'S FAVORITE SEASON

- More than 250 million people worldwide suffer with asthma, in which the lung's tubular airways become constricted and inflamed.
- Fall is the prime season for flareups for a variety of reasons, including rapid temperature changes and breezes that carry pollen and mold spores. Schoolchildren are especially vulnerable.
- Asthma lasts a lifetime, and like the common cold, it is incurable. But it is also controllable.

+ + +

Bill Clinton, John F. Kennedy, and at least four other U.S. presidents; the Reverend Jesse Jackson; composers Ludwig von Beethoven and Antonio Vivaldi; authors Charles Dickens and Edith Wharton; Olympic medalists Jackie Joyner-Kersee, Jim Ryun, and Greg Louganis; NBA stars Isaiah Thomas and Dennis Rodman; and actress Elizabeth Taylor.[1] They all had something in common with more than 250 million people on the planet today, including this author and two sons.

We all live with the whims and torments of asthma, an ailment that can make the fresh air, so important to our well-being, mutate into a health hazard. And by all accounts, our legions are growing. The American Lung Association has estimated that as of 2018, 42 million of those identified as having asthma were living just in the United States, which would represent a 43 percent jump over 1990.[2] However, you will see different figures out there from other credible health sources, and any estimate you see would be imprecise. Misdiagnosis is an issue, and countless numbers of sufferers never make the spreadsheets, as noted by a group of British researchers: "It appears that a substantial proportion of patients simply don't ever report their symptoms to a doctor."[3]

Even accounting for diagnostic inconsistencies, the evident increase in cases is significant, and in many ways mysterious. Why is this happening now? Could it be pollution? Actually, air quality unquestionably has improved in this country.[4]

In fact, it has improved so much in the developed world over the last 40 years that atmospheric scientists are confident that the reduced aerosols are allowing more sunlight to warm the North Atlantic Ocean and, thus, contributing to more tropical storms and hurricanes.[5]

Whatever is behind the surge, school nurses, who are on the battlefronts, can tell you a whole lot about asthma, a disease that they and medical researchers still are trying to understand more fully. One thing they've learned for certain, said Kate King, who works in the Columbus, Ohio, school district and is head of the National Association of School Nurses, is that they don't know everything. They keep learning. For example, she said, "We used to think that children outgrew asthma. We know it's not really true. Once you have asthma you always have asthma."[6]

WHAT IS ASTHMA?

For those with asthma. the tubular airways that import and export air to and from the lungs periodically become inflamed and narrower. This can make the normally simple act of breathing difficult and painful, provoking coughing and wheezing.[7]

Attacks can be set off by any number of factors, including allergies, exercise, stress, and weather, and asthma can interact with background pulmonary diseases. Asthma can be of the "intermittent" or the flare-up type, or "persistent," in which symptoms are chronic. It can be "adult onset" or "pediatric."[8]

Researchers have found evidence of correlations between some circumstances and asthma prevalence, and I can personally attest to at least two of those identified by the World Health Organization. Asthma,

WHO says, is a family affair, and it is more likely in those who have hay fever.[9] Those qualifications would apply to my sons and me.

However, so much about asthma remains an enigma. It is a complicated condition and likely encompasses "many diseases," Pennsylvania respiratory specialist Dr. Ruthven Wodell, who has been practicing for more than 40 years, told me. As our family allergist, I've had many conversations with Dr. Wodell about respiratory issues over the years, and his treatments have improved the quality of my life.

I have what I hold to be a healthy skepticism of medications, but I can vouch for steroid inhalers. I've had nights when I thought my airways were being squeezed to the point of closing, akin to a cramping sensation. When I used the steroid inhalers as prescribed—as in doing one or two puffs regularly, not just at the point of discomfort—the symptoms backed off. I have to believe that was more than coincidence. But my discomforts are petty compared to what others have suffered, including our older son. For him, one particular autumnal asthma attack was a life crisis. Fortunately, he was able to get to an emergency room in time.

Sufferers who experienced serious issues and their family members have shared compelling stories on the Lung Association website and elsewhere. A New York woman told of how she and her brother were born with asthma; how she developed bronchitis and her brother, pneumonia, at a young age; and how their asthma flared when atmospheric moisture increased. A Florida father related how his nine-year-old son had died as a result of an asthma attack.[10]

A 71-year-old woman discussed the frustrations and physical discomforts she experienced when visiting her first great-grandchild in Arizona. She had endured some weather-related flare-ups on the way. The house was another challenge: The family had two dogs and four cats, and the pet dander was a serious asthma trigger. She couldn't stay there, resulting in "hurt feelings all around, which became criticism and judgment of me and my asthma. . . . It's such a hard place to be."[11]

AUTUMNAL BLITZ

With so many potential triggers and demographic targets, asthma is a year-round affliction, but autumn appears to be the harvest season for attacks. A nurse at an urban school where about 10 percent of the pupils suffered with asthma, told me that almost at the stroke of autumn, her office will swell with the sounds of coughing and wheezing.[12]

A study of "unexplained" increases in asthma-related hospital admissions documented that they peaked in the third week in September and didn't tail off until later in the year. The researchers noted that the great Hippocrates recognized the correlation between autumn and the bump in asthma symptoms. They referred to analysis that documented that fully 50 percent of all asthma-related hospital admissions in the city in the 1969–1971 period among 5- to 34-year-olds occurred in September and October. A Canadian study found a 60 percent increase in asthma mortality in those months during the 1975–1981 period.[13]

The explanations for an autumnal spike would be manifold. Directly and indirectly, the behavior of the atmosphere is a prominent player.

Temperature Changes

Like spring, fall is a transition season characterized by volatility. In the temperate zones, the ambitious winter encounters stiff resistance from the reluctant summer. The result can be radical changes in temperature that can exacerbate asthma conditions. Those with respiratory conditions, especially children who typically are asthma-free during the warmth of the summer, can get caught in the crossfire.[14]

Thunderstorms

In the atmospheric battleground, cold fronts are inevitable, and autumn is a time of brisk frontal traffic that can lead not only to rapid temperature change but ignite thunderstorms in temperate areas.

Cold fronts move quickly, at about double the speed of warm fronts. The denser, cooler air arrives in a rush, forcing the lighter, warmer air in its path to rise, and can set off potent storms. With the passage of the front come temperature and pressure changes and gusty winds.[15]

Those sometimes-violent storms can be dangerous for asthmatics, and frontal passages aren't the only sources. In the populous eastern half of the country, fall is a season for tropical-storm remnants that historically have spawned powerful thunderstorms that generate cold downdrafts.

These rapidly descending currents can be loaded with particulates including pollen, which can be especially troublesome for asthmatics.[16]

Under certain circumstances, thunderstorm downbursts can even create dust storms.[17] Strong thunderstorms were blamed for a rash of asthma attacks that sent 8,000 people to emergency rooms in the fall of 2016 in Melbourne, Australia.[18]

Pollen

Pollen allergies are well-known triggers, and the ragweed season laps deeply into fall in some of the nation's most populous regions. Ragweed is tenacious and ubiquitous, and is a prodigious pollen producer. It needs neither care nor love. In fact, it appears to thrive on hate and disdain. It can grow in pastoral settings, or on jungled lots, or streets, or driveways. One single homely plant can send a billion pollen grains airborne.[19] As New Yorkers learned (see chapter 1), efforts at eradication can be humorous.

Seeds that lay dormant in the ground for decades can grow under the right circumstances. Ragweed lives happily in the countryside and on the urban heat islands. One analysis found that pollen counts in a city can be seven times higher than in the surrounding countryside,[20] a potentially troublesome imbalance for people of low income.

I have had my share of exhausting ragweed sneezing attacks that left me dazed, useless, and amazingly hungry, with a particular craving for sweets. I have yet to find a good explanation for what all that sneezing had to do with appetite and a self-destructive desire for sugar. Then again, based on what I've learned about seasonal-change symptoms, perhaps the autumnal loss of daylight had more to do with appetite than does ragweed. In any event, I did relearn that life holds worse sensations than satisfying sugar cravings.

Mold Spores

The mold spore may well be the ugly duckling of the allergen family, but allergists have insisted that it is greatly underrated as a tormentor and ranks with the common cold for garnering disrespect.

Dr. Donald Dvorin, an asthma and allergy specialist who practices in New Jersey and has been with the National Allergy Bureau for more than 30 years, said that about 60 percent of the patients in his former practice who were allergic to grass, tree, and ragweed pollen, also reacted to mold spores.[21]

Autumn is the heyday of the airborne mold spore—separate from indoor mold, which is another hazard altogether. Damp leaves and rotting logs are more than willing hosts for the growth of outdoor molds that are only too happy to send their spores into the air. Ironically, the wetness and dryness of autumn both work against the allergic. Rain promotes favorable conditions for mold growth. When it's not raining, the seasonally decreasing moisture level of the atmosphere—commonly called humidity—makes the air more conducive for the flights of the spores. Spores can also spread on the cat-like feet of fog.[22]

THE NOT-SO-GREAT INDOORS

As temperatures fall, the weather drives more people indoors and keeps them inside longer, and germs couldn't be more grateful for opportunities to enter unsuspecting hosts.[23] Unfortunately, sometimes those hosts are asthmatics.

INHALER SEASON

For evidence of asthma's prevalence in autumn, look no further than the nation's schools. Kate King doesn't even have to look. She lives it. Along with running the Nurses Association, she is the school nurse at the World Language Middle School in Columbus. The student body is about 94 percent people of color, and majority Latino.[24]

Among schoolchildren, asthma is a leading cause of school absenteeism, according to King's association. It disproportionately affects inner-city children, low-income people, and people of color, all of whom experience more asthma-related emergency-room visits than the general population.[25]

A lot of students in King's school have asthma issues, and fall is prime time. While the weather and other multiple factors likely are driving the increase in autumn flare-ups, King advises against ignoring the obvious.

The students are back in school, literally back in touch with each other. "We have this captive audience," she said. "Kids are coming together. Kids who have asthma are getting colds, flu, other respiratory diseases."

The war on asthma has become part of her daily routine, and she is well-armed with an arsenal that includes rescue inhalers. "It's work up front," she said. "I give them their daily medicine every morning." That can prevent "an acute asthma attack later in the day. It only takes about three minutes to do their inhaler. In the afternoon, if I got someone having an asthma attack that can be an hour or an hour and a half." And, yes, she said, those attacks can be life-endangering.[26]

Sometimes, the worst happens.

FATAL CONSEQUENCES

One study documented that 38 schoolchildren died from asthma attacks from 1990 to 2003. It found that the majority of the fatalities occurred among teenagers, and a disproportionate number—18—of the victims were Black students. In 16 of the cases, the children were engaged in physical activity, and 12 died while they were awaiting medical assistance.[27]

King recalled a well-publicized incident in Philadelphia, which occurred in 2013. On September 25 of that year, 12-year-old Laporshia Massey died from an asthma attack she suffered at a school that did not have a nurse on duty. According to a suit filed by her estate, Laporshia told her teacher she was having trouble breathing but was ordered to "be calm." Although she was clearly in the throes of a medical crisis, no one summoned an ambulance. After arriving home, she would die from respiratory failure on her way to a hospital.[28]

In making an argument for increasing the ranks of school nurses in the district, Jerry Jordan, then head of the Philadelphia teachers union, invoked the 2013 tragedy, calling it "a devastating tragedy that cannot ever be forgotten."[29]

Inhalers have become standard equipment in many schools, and the fact that they are in Ohio has a lot to do with Kate King, who in 2021 was named the "School Nurse Administrator of the Year" by the Ohio Association of School Nurses.[30] It was around the time of that death in Philadelphia that King began lobbying for the Ohio law that required schools to have inhalers at the ready. "We were seeing so many kids with asthma problems," she said.

Early-life issues, such as premature birth, low birth weight, viral infections, and exposure to tobacco and other pollutants can increase asthma risk, states the WHO.[31]

As to why asthma has been so prevalent in urban areas, King and the Nurses Association have been examining the whole menu of possibilities, including any role in worldwide warming. "We need to look at all those things," she said.[32]

TOO CLEAN?

The "hygiene hypothesis" that is a suspect in the increase of pollen allergies, as discussed in chapter 1, could well be one of those factors. Some studies have concluded that perhaps the presence of farm animals provides beneficial germ exposure—that people who live on farm properties tend to have fewer allergies.[33]

The hygiene hypothesis suggests that during those critical several weeks after birth, when the child's body is learning to defend itself, the immune response is "derailed" in the clean household environments in the developed world. "In other words, the young child's environment can be 'too clean' to pose an effective challenge to a maturing immune system," according to the U.S. Food and Drug Administration (FDA). The body's defenses "end up being so inadequate that they actually contribute to the development of asthma." The FDA notes that the hypothesis is supported by epidemiologic studies.[34]

This gives considerable credence to the hygiene hypothesis, a term whose origin has been credited to a 1989 paper written by the University of London's David Strachan, who based his findings on 17,000 case studies.[35] (In a paper he coauthored in 2022, Strachan noted that the term actually had been kicking around since 1985.[36])

The hygiene hypothesis has had considerable support in the years since. "During its 30 years history, the Hygiene Hypothesis has shown itself to be adaptable whenever it has been challenged by new scientific developments," a team of German researchers wrote in a 2021 study funded by the German Center for Lung Research. The researchers concluded that the hypothesis was a factor in at least some asthma cases. And while acknowledging that their discussion was "brief," they suggested that the hygiene hypothesis might hold promise for insights into cancer and psychiatric disorders, and "a more generalized explanation for health burden in highly industrialized countries."[37]

ANTIBIOTICS . . . AND TYLENOL?

The matrix of possible causes of the disturbing growth in asthma incidence appears to be quite complex, with a legion of suspects. Disentangling the roles of the atmosphere and other environmental factors from other possible causes remains an international work in progress.

For example, the increases in asthma diagnoses have coincided with rising usage of antibiotics in children, and while findings conflict—sound familiar?—some research has documented that the relationship is more than a mere correlation. It appears that antibiotic use early in life alters bacteria in such a way that conditions can become favorable for development of asthma.[38] "Antibiotics, among the most used medications in children, affect gut microbiome communities and metabolic functions," medical researchers from New York and Rutgers Universities and the University of Zurich said in a 2022 paper in *Mucosal Immunology*.[39]

In plain English, senior author Martin Blaser, director of the Center for Advanced Biotechnology and Medicine at Rutgers, commented: "The practical implication is simple: Avoid antibiotic use in young children whenever you can because it may elevate the risk of significant, long-term problems with allergy and/or asthma."[40]

Another suspect is acetaminophen, commonly sold under the brand name Tylenol, a popular pain remedy for children. Research has returned mixed verdicts on the possible connections, but "several epidemiologic observations suggest that acetaminophen use might be a risk factor for asthma development, as well as asthma exacerbation," according to doctors with the College of Family Physicians of Canada. Tylenol may cause "airway inflammation, bronchoconstriction, and subsequent symptoms of asthma."[41]

OBESITY

"Obesity epidemic" has become a national cliché, but in this case, it turns out that even clichés can be true. The evidence clearly indicts obesity as a major contributor to the swelling asthma numbers.

Based on data from the Centers for Disease Control and Prevention, asthma cases are 23 percent higher among the obese, compared with the rest of the population.[42]

About two in five Americans meet the body mass index criterion for obesity, according to the National Institutes of Health.[43] Based on the U.S. Census numbers for the adult population,[44] up to 10 million obese adults may be asthmatics. NIH estimates that 250,000 new asthma cases per year in the United States have connections with obesity. This relationship has radically changed the demographics of asthma in this country. The prevalence of asthma in lean adults is 7.1 percent, compared with 11.1 percent among the obese. The imbalance is more striking in women—the prevalence of asthma in lean versus obese women is 7.9 percent and 14.6 percent, respectively.[45]

What does body fat have to do with asthma?

Extra weight near the chest and abdomen might constrict the lungs, says the American Lung Association. Fatty tissue can produce "inflammatory substances that might affect the lungs." What's more, the association's own researchers found that medications don't work as well for the obese. Pulmonary medicine specialists at the University of Vermont and University of Pittsburgh concurred. In addition, they said in 2018, "obese asthmatics have more symptoms" and "more frequent and severe exacerbations." They also found that "Obese children tend to have increased asthma severity," and one statistic they cited I found astounding: "Almost three in five adults in the United States with severe asthma are obese."[46]

The obese are more likely to have other medical problems—such as depression and sleep apnea—that can intensify asthma symptoms.[47] And in the surges in both obesity and asthma, medical experts see a common bond—lifestyle changes, and something missing from the lives of those afflicted with those conditions. "Obese asthmatics exercise less," observed the authors of a 2022 article in the journal *Nutrients*. They spend less times outdoors, thus depriving themselves from the most important natural source of vitamin D—the sun. That may make them more sus-ceptible to respiratory infections,[48] about the last things asthmatics need.

VITAMIN D AND ASTHMA

Diets and sedentary behavior are obvious drivers of obesity, and thus, asthma proliferation. Another important driver may be vitamin D—that "sunshine vitamin" so essential to the lungs and the immune system. Some experts believe rising rates of asthma may, in part, be due to a defi-ciency of vitamin D. Vitamin D is essential for lung and immune system development. Kids these days are spending less time outside, depriving themselves of an important source of vitamin D—sunlight.[49] We will discuss vitamin D further in the winter section.

Despite the assortment of weather triggers, it's not a stretch to say that if kids spent more times outside, not only would it be a blow against asthma, it might take some of the air out of the obesity epidemic.

If you are diagnosed with asthma, here are a few practical tips from the World Health Organization.

- **Know your symptoms.** Pay attention to them, including cough-ing, wheezing, and difficulty breathing, for signs that symptoms are worsening.
- **Know your triggers.** Be they weather changes, smoke, pollen, animal fur, or certain fragrances, try your best to stay out of harm's way. If that's not possible, keep a reliever inhaler available.
- **Know your inhalers.** A bronchodilator opens up the small air-ways. A steroid inhaler reduces inflammation of the lungs. Be aware of your symptoms.
- **Take control.** Ask your doctor how the inhaled medications work, and make sure that family and friends know what to do if your asthma flares up.[50]

ADDENDUM

"DON'T SHUT THE WINDOW"

In their own words, courtesy of the National Archives, here are John Adams's and Benjamin Franklin's thoughts on how one might catch a cold. Keep in mind that this was a good 100 years before germ theory became widely accepted. The first excerpt is from John Adams's diary;[1] the second is a letter from Franklin to Dr. Benjamin Rush.[2] The original syntax and spellings, as they appeared in the Archives, have been retained.

+ + +

Monday September 9, 1776

On this day, Mr. Franklin, Mr. Edward Rutledge and Mr. John Adams proceeded on their Journey to Lord Howe on Staten Island, the two former in Chairs and the last on Horseback; the first night We lodged at an Inn, in New Brunswick. On the Road and at all the public Houses, We saw such Numbers of Officers and Soldiers, straggling and loytering, as gave me at least, but a poor Opinion of the Discipline of our forces and excited as much indignation as anxiety. Such thoughtless dissipation at a time so critical, was not calculated to inspire very sanguine hopes or give great Courage to Ambassadors: I was nevertheless determined that it should not dishearten me. I saw that We must and had no doubt but We should be chastised into order in time.

The Taverns were so full We could with difficulty obtain Entertainment. At Brunswick, but one bed could be procured for Dr. Franklin and me, in a Chamber little larger than the bed, without a Chimney and with only one small Window. The Window was open, and I, who was an invalid and afraid of the Air in the night, shut it close. Oh! says Franklin dont shut the Window. We shall be suffocated. I answered I was afraid of the Evening Air. Dr. Franklin replied, the Air within this Chamber will soon be, and indeed is now worse than that without Doors: come! open the Window and come to bed, and I will convince you: I believe you are not acquainted with my Theory of Colds. Opening the Window and leaping into Bed, I said I had read his Letters to Dr. Cooper in which he had advanced, that Nobody ever got cold by going into a cold Church, or any other cold Air: but the Theory was so little consistent with my experience, that I thought it a Paradox: However I had so much curiosity to hear his reasons, that I would run the risque of a cold. The Doctor then began an harrangue, upon Air and cold and Respiration and Perspiration, with which I was so much amused that I soon fell asleep, and left him and his Philosophy together: but I believe they were equally sound and insensible, within a few minutes after me, for the last Words I heard were pronounced as if he was more than half asleep. . . . I remember little of the Lecture, except, that the human Body, by Respiration and Perspiration, destroys a gallon of Air in a minute: that two such Persons, as were now in that Chamber, would consume all the Air in it, in an hour or two: that by breathing over again the matter thrown off, by the Lungs and the Skin, We should imbibe the real Cause of Colds, not from abroad but from within. I am not inclined to introduce here a dissertation on this Subject. There is much Truth I believe, in some things he advanced: but they warrant not the assertion that a Cold is never taken from cold air. I have often conversed with him since on the same subject: and I believe with him that Colds are often taken in foul Air, in close Rooms: but they are often taken from cold Air, abroad too. I have often asked him, whether a Person heated with Exercise, going suddenly into cold Air, or standing still in a current of it, might not have his Pores suddenly contracted, his Perspiration stopped, and that matter thrown into the Circulations or cast upon the Lungs which he acknowledged was the Cause of Colds. To this he never could give me a satisfactory Answer. And I have heard that in the Opinion of his own able Physician Dr. Jones he fell a Sacrifice at last, not to the Stone but to his own Theory; having caught the violent Cold, which finally choaked him, by sitting for some hours at a Window, with the cool Air blowing upon him.

+ + +

Dear Sir,

I received your Favour of May 1. with the Pamphlet for which I am obliged to you. It is well written. I hope in time that the Friends to Liberty and Humanity will get the better of a Practice that has so long disgrac'd our Nation and Religion.

A few Days after I receiv'd your Packet for M. Dubourg, I had an Opportunity of forwarding it to him by M. Poissonnier, a Physician of Paris, who kindly undertook to deliver it. M. Dubourg has been translating my Book into French. It is nearly printed, and he tells me he purposes a Copy for you.

I shall communicate your judicious Remark relating to Air transpir'd by Patients in putrid Diseases to my Friend Dr. Priestly. I hope that after having discover'd the Benefit of fresh and cool Air apply'd to the Sick, People will begin to suspect that possibly it may do no Harm to the Well. I have not seen Dr. Cullen's Book: But am glad to hear that he speaks of Catarrhs or Colds by Contagion. I have long been satisfy'd from Observation, that besides the general Colds now termed Influenza's, which may possibly spread by Contagion as well as by a particular Quality of the Air, People often catch Cold from one another when shut up together in small close Rooms, Coaches, &c. and when sitting near and conversing so as to breathe in each others Transpiration, the Disorder being in a certain State. I think too that it is the frowzy corrupt Air from animal Substances, and the perspired Matter from our Bodies, which, being long confin'd in Beds not lately used, and Clothes not lately worne, and Books long shut up in close Rooms, obtains that kind of Putridity which infects us, and occasions the Colds observed upon sleeping in, wearing, or turning over, such Beds, Clothes or Books, and not their Coldness or Dampness. From these Causes, but more from too full Living with too little Exercise, proceed in my Opinion most of the Disorders which for 100 Years past the English have called Colds. As to Dr. Cullen's Cold or Catarrh à frigore, I question whether such an one ever existed. Travelling in our severe Winters, I have suffered Cold sometimes to an Extremity only short of Freezing, but this did not make me catch Cold. And for Moisture, I have been in the River every Evening two or three Hours for a Fortnight together, when one would suppose I might imbibe enough of it to take Cold if Humidity could give it; but no such Effect followed: Boys never get Cold by Swimming. Nor are People at Sea, or who live at Bermudas, or St. Helena, where the Air must be ever moist, from the Dashing and Breaking of Waves against their Rocks on all sides, more subject to Colds than those who inhabit Parts of a Continent where the Air is dryest. Dampness may indeed assist in producing Putridity, and those Miasms which infect us with the Disorder we call a Cold, but of itself can never by a little Addition of Moisture hurt a Body filled with watry Fluids from Head to foot.

I hope our Friend's Marriage will prove a happy one. Mr. and Mrs. West complain that they never hear from him. Perhaps I have as much reason to complain of him. But I forgive him because I often need the same kind of Forgiveness. With great Esteem and sincere Wishes for your Welfare, I am, Sir, Your most obedient humble Servant.

B Franklin

WINTER

THE ICING ON THE CALENDAR

In that mythical poll we invoked in the spring section prelude, winter in all likelihood would finish last in any ranking of seasons in terms of promoting well-being.

I would argue that winter would deserve more respect from the electorate.

For one thing, in colder climates, winter is the mortal enemy of the deadliest animal on earth. (Can you guess what it is? Hint: It's not a lion or snake. The answer is in chapter 11.)

Our winter addendum, and the experiences of masses who have endured Arctic outbreaks through the years notwithstanding, the prospect of milder winters in a warming planet has become a public health concern.

Among generally healthy people, the body does an excellent job of adapting to the cold. One important strategy is more or less the reverse of sweating: the shivering response, which generates heat to warm up the body. (An aside: With more surface area to heat, taller people tend to get colder faster than their shorter counterparts. That's all other things being equal.) We have an excellent insulator. That would be our fat, providing it's not all in the gut.[1]

Winter doesn't have to be an invitation to isolation and lethargy, even in the coldest places. It offers ample opportunity for outdoor exercise, and not just skiing and ice-skating. For those in decent shape with sound cardiovascular systems, even snow-shoveling can be an excellent workout.

And bright sun atop a snow cover—a personal favorite—is a natural encounter with light therapy, an antidote to the "winter blues."

Yes, winter has dark sides that probably would outdo spring's. Snow and ice, of course, have been responsible for more than one visit to an emergency room after a bone-breaking or concussive fall.

Snow-shoveling is not for everyone. An American Heart Association study identified it as being particularly hazardous to those who are out of shape. Even a minor bit of shoveling can put as much strain on the heart as a treadmill stress test, according to the study's lead author, cardiologist Dr. Barry Franklin, with Oakland University in Michigan. Shoveling is primarily arm work, "which is more taxing and demanding on the heart than leg work." In addition, if you're shoveling, chances are it's cold outside, and exposure can constrict arteries that "are about the size of cooked spaghetti." Shoveling results in "hundreds of deaths" annually, said Franklin.[2]

While those fatality figures are disturbing, they aren't in a league with the deaths attributed annually to influenza, and that is just in the United States. Winter is decidedly the prime flu season. People spending so much time inside in the cool seasons often has been cited as a leading cause of the spread of influenza. That may well be a factor, but as a team of researchers pointed out, it isn't as though people stop congregating when the weather gets warmer. Why aren't outbreaks more frequent in summer when people continue to report to their workplaces, and travel on crowded airplanes, or on cruise ships?[3]

What the atmospheric connections might be remains one of those innumerable research questions. What is clear is that influenza has definitive seasonal rhythms, that it is primarily a wintertime affliction wrought by a "seasonal stimulus," in the words of Dr. Robert Edgar Hope-Simpson, a British physician.[4]

One plausible hypothesis, explored in chapter 12, is the loss of exposure to sunlight, the most significant source of natural vitamin D. Deficiencies of vitamin D in winter might have something to do with flu seasonality.

A less serious affliction associated with winter is "cabin fever." Like spring fever, it has no official medical definition. However, it is real, and evidently a whole lot of people experienced it during the COVID-19 lockdowns. For a summary of the symptoms, I'll defer to the one offered by a specialist at Britain's Institute of Mental Health, who described it as "commonly understood to refer to a combination of anxiety, lassitude, irritability, moodiness, boredom, depression, or feeling of dissatisfaction in response to confinement."[5]

When the term first appeared is unclear, but its origin is credited to B. M. Bower, who wrote a Western-themed novel, *Cabin Fever*, published in 1918. "Just as the body fed too long upon meat becomes a prey to that horrid disease called scurvy," she wrote, "the mind fed too long on monotony succumbs to the insidious ailment that the West calls 'cabin fever.'"[6]

Major winter storms and extreme cold can, on occasion, deter people from spending time outside and lead to cabin-fever symptoms. Fortunately, for most people, under most circumstances, it is a very temporary condition. Experts say it can be alleviated by simple measures such as exercising regularly, even if indoors; staying in touch with others in the outside world; taking time to relax; spending at least some time in the winter air; and indulging in some creative activity.

Paul Siple and Charles Passel are two people who in all likelihood had more exposure to cabin fever than most people on the planet. They were members of Admiral Byrd's expeditionary team in Antarctica, and based on Siple's diary, their daily lives redefined "tedium." But they did engage in creative activity, and a project they undertook changed the way we experience the cold of winter. They came up with an index to characterize how humans might feel in given wind and temperature conditions.

They called it "wind chill."

CHAPTER 10

COLD REALITIES

- From the days of Hannibal's march across the Alps to the twenty-first century, extreme cold has been one of the most dangerous and deadliest weather phenomena.
- The planet is getting warmer, but meteorologists assure us that we can expect winter cold shots to continue for the foreseeable future.
- For what we know about how cold affects the body, we can thank Antarctic explorers, Canadian and U.S. researchers, and 12 volunteers who risked their faces for science.

+ + +

When I stepped off the plane, what greeted me was something more profound than a chill, an invisible force that penetrated to the marrow, a depth of cold that I never had experienced before. The wind wasn't stinging so much as numbing. My being was consumed by an instant shivering sensation—shivering like a human paint-mixer, actually—which I know now was only my body's way of trying to save my life.

It's not as though we had just arrived in Montreal from Miami: When my wife-to-be and I left Philadelphia that morning in mid-February, the temperature was in the 20s Fahrenheit range. Now we were in the land of the born-tougher Quebecois, not among the weather wusses of the Mid-Atlantic who cower at the mention of "wind chill."

We were in the land of wind chill, in the country where the research community, with assists from the United States, faced the challenge of answering the "how does it feel" question. What resulted at the turn of the millennium was a measure that aimed to capture how the wind and cold conspired to challenge our bodies. The scale these Canadians developed is commonly accepted today, even if most people have no idea where it came from.

This measure will remain useful and retain its value for the foreseeable future. Even though the world is getting warmer, meteorologists assure us that we are not done with cold shots.

YOUR MOTHER'S SWEATER

At a biometeorology conference in Kansas City some years ago, I heard Randall Osczevski offer a Canadian perspective on the cold. Osczevski, a retired physicist who worked for the Canadian defense department and is an international—and even an interplanetary—authority on how cold affects the body, remarked: "Someone once said that a sweater is something you put on when your mother is cold."[1]

Maybe in Canada.

On that particular February in Montreal I didn't feel quite as weather-wussy when the doorman at our hotel greeted us by saying, "You picked a cold weekend to come here." I'm figuring, and this guy actually lives here. Besides, my fiancée, a McGill University graduate who endured four Montreal winters, acknowledged that it was, in fact, cold.

On the TV in our room, the Weather Channel was carrying dire cold warnings. The forecast for the next day called for a "high" of 7 degrees below zero Fahrenheit, after a low of 24 below, which would be a record for the date—in Montreal. If it ever got that cold in Philadelphia, where it's been more than 30 years since the last time it hit zero, the entire region would have been shut down indefinitely.

Yet, we managed to survive. We even ventured outside and, with Montrealers as our role models, did the things they did: Walked, browsed in shops, more or less behaving as though this was just a typical winter weekend. We waited outside for 45 minutes in a restaurant line along with a variety of hatless Montrealers.

My wife even swam in an outdoor pool—heated, of course, with a steamy top that gave the effect of a massive cup of cappuccino. Despite our questionable judgment, somehow our bodies adapted. Had we been aware of how hard our bodies were working, we would have been more than impressed and grateful. Perhaps we were foolish; overexposure to the cold unquestionably can be life-threatening.

For developing deeper understanding of just what our bodies can tolerate, we can thank Randall Osczevski, the late Maurice Bluestein, other researchers, and 12 unsung heroes who risked their noses, cheeks, and foreheads for the sake of science.

OUR INTERNAL FURNACE

In chapter 4, we talked about the body's ingenious response to the heat. Its response to cold is a tribute to the body's versatility and that afore-mentioned heroic hypothalamus, which is a huge player in regulating our adaptation response.

A healthy body temperature is considered to be about 98.6 degrees Fahrenheit, give or take a few ticks. It is not precise, because technically it refers to core temperature of "internal vital organs," but, thankfully, the readings typically are taken in more accessible places, commonly the mouth. The temperature fluctuates, depending on what we are doing, or not doing. It is higher in the afternoon than in the morning.

When our temperature deviates significantly from the averages, the hypothalamus goes to town. If we're hot, it will order up sweat for cool-ing; if the temperature is too low, it stokes the furnace.[2]

The Shakes

It was perfectly understandable that the first thing I did when we set foot on Montreal that weekend was shiver. Shivering can be viewed as more or less the winter counterpart to sweating in summer.

It was our body's way of fighting back by trying to generate heat. While shivering, the muscles alternately expand and contract, warming the body.[3]

Cold Hands, Warm Heart

Ever wonder why your fingers can feel like they are about to turn to icicles even though you are layered to the max under two sweaters and are wearing double-insulated gloves?

The blood coursing through our arteries and veins acts like a heating fluid that keeps our hands and feet warm. But in the cold, the body has priorities. It will do what it can to warm the heart and brain and other important organs. No offense is meant to the extremities, but the hands and feet pay the price as their share of the blood supply is rationed.[4]

HOW MUCH CAN WE TAKE?

Even the mighty hypothalamus has limits. If the body's core temperature falls below about 95 degrees Fahrenheit, and the heat loss outpaces heat production, the body enters the dangerous state of hypothermia. The groups most susceptible to this potentially life-threatening condition include older people without access to proper food and clothing. However, otherwise-healthy people who spend significant time outdoors, such as hikers and runners, are by no means immune.[5]

Hypothermia evidently was a factor in the 2021 ultramarathon tragedy in China, in which 21 participants died when hail and high winds struck a stretch of mountainous trail.[6]

And it doesn't have to be all that cold. For ultrarunners, hypothermia can be a threat even in summer.[7]

ABOUT HYPOTHERMIA

Among other defining traits, we human beings are "homeotherms." As environmental temperatures change, so do our body temperatures. As researchers noted in a 2022 article published in the journal *Physiology*, accelerating global warming may pose new threats to public health, but the human body has been challenged to adapt to climate change from the beginning. History offers numerous examples of the killing powers of prolonged exposure to cold, from the deaths that occurred when Hannibal's army crossed the Alps, to the fatal consequences of Adolf Hitler's invasion of frigid Russia during World War II. Autopsy records aren't necessary to conclude that in all likelihood, hypothermia has been at least a contributing factor in millions upon millions upon millions of deaths throughout human history.[8]

Part of the reason for the cold's lethal power is that one of the organs affected when the body temperature drops is the brain. The victim may be unaware of what is happening.[9] The symptoms of hypothermia tend to begin gradually, and the resulting mental confusion may result in victims of hypothermia making risky decisions that further jeopardize their lives.[10]

It is estimated that half of the older people afflicted with hypothermia die either before or soon after they are found.[11] Victims might not realize that the wise decision would be to seek emergency treatment.

What's more, a person with hypothermia also is likely to be suffering from frostbite.

ABOUT FROSTBITE

While not in the life-threatening league with hypothermia, frostbite also is a dangerous condition that you absolutely want to avoid. It can penetrate to deep layers of skin and muscles, and even can damage blood vessels, cutting off flow to the wounded area and leading to gangrene. That can result in the amputation of a limb.[12]

"Frostbite" is one of those words that tidily captures what it is. While the term appears to date to the 1500s,[13] in all probability it began when the first humans had the pleasure of lengthy exposure to the cold without adequate protection. The earliest verified evidence of frostbite is believed to have been discovered on the body of a 5,000-year-old mummy found in the Andes Mountains.

Napoleon's chief surgeon reported massive numbers of cold-related injuries during France's ill-fated invasion of Russia in the winter of 1812–1813. As Germany would learn later, Russia isn't a winter playground.[14]

Frostbite results when ice crystals form in skin tissues, causing breaks. In more serious cases, blisters form that fill with fluid. In "third degree" frostbite, the freezing can penetrate to deep layers of skin and muscle.[15]

Take the case of Fahad Badar, a Qatari banker, who had experienced about every cold-weather challenge known to human life: He is one of the world's most famous mountain climbers. His conquests include the legendary Mount Kilimanjaro, the highest mountain in Africa. He also suffered one of the world's most famous cases of frostbite, which he developed during an expedition to Broad Peak, in Pakistan, in July 2020. Seven months later, Badar posted a haunting TikTok video of his frozen, blackened fingers—which he was about to lose to amputation. The video was seen by 45 million viewers.[16] One would hope that his story prevented at least some of the viewers from making similar miscalculations.

It makes intuitive sense that hypothermia and frostbite would have a well-developed and perilous relationship. Because people who develop hypothermia gradually lose their ability to make judgments, and their mobility, they may be overtaken by drowsiness and confusion.[17]

Precisely what conditions can induce hypothermia and/or frostbite ultimately depends on the bodies and circumstances in question. However, some researchers have taken extraordinary measures to identify the boundaries between what our bodies can endure and what may put us on the edges of injury and, in some cases, mortal danger.[18]

HOW DOES IT FEEL, WINTER VERSION

High on that list of researchers would be Osczevski and Maurice Bluestein, a mechanical engineer and professor at Purdue University. For Bluestein, his interest in what happens when the cold meets the skin dates to January 1994, a month that was characterized by Arctic cold in Indianapolis, where the temperature got down as far as 27 degrees below zero Fahrenheit.[19]

One evening during that month, Bluestein was shoveling snow to liberate his daughter's car. The National Weather Service warned that the wind chill was a perilous 65 degrees below zero, so cold that it could cause frostbite almost instantly. Bluestein nonetheless went outside to pursue his fatherly mission. But something strange happened as he proceeded to shovel, according to his obituary in the *New York Times*. He was so warm, he found himself peeling off some of his layered clothing. Worse, "He was sweating," recalled his daughter. Something had to be wrong with that wind-chill scale—"This makes no sense." Later, at a conference in Toronto, Bluestein happened to meet Osczevski, a physicist with the Canadian defense department, who likewise was perplexed by what he viewed as inaccurate wind-chill values.[20]

The numbers were derived from an experiment by Paul Siple and Charles Passel during a U.S. Antarctic expedition right before World War II, in which they suspended a bottle of water from the roof of a building and measured the container's loss of heat as the water froze.[21] We will get back to that experiment—which Osczevski and Bluestein identified as the most memorable result of the Byrd expeditions, and which would not be a unanimous opinion—and to Siple and Passel and what they were doing in Antarctica.

Relevant to the Siple–Passel exercise, Osczevski constructed an artificial human head as part of an exercise to measure heat loss at the average height of the human face—a face value, as it were.[22] As it turned out, in experiments, Osczevski and Bluestein found that the heat loss per unit area of the face wasn't "very different" from the heat loss from the Siple–Passel container. That explained why the index definitely had some value. But they were among those who believed the index overvalued the effects of the wind, and they had plenty of expert company in their skepticism.

Osczevski and Bluestein were among the participants in an internet conference on wind chill convened by the Canadian Weather Service in April 2000. It was a massive working session with 400 participants from 35 countries. If the attendees agreed on anything, it was that the world needed a new index to explain how the cold felt.[23] It was a complex undertaking that remains a work in progress, for as Osczevski and Bluestein observed, "wind chill is not a neat and simple package."[24]

Ultimately, the two of them were assigned to take on the problem and generate a new index. Working together and exploiting advances in the science and computer modeling, they developed a more accurate scale for assessing how wind and cold affects the body.[25]

Some of the new values, calculated for a range of temperatures from 40 degrees Fahrenheit to 45 degrees below zero, and winds from 5 to 60 miles per hour, were radically different from the old. For example, in the old chart that was in place when Bluestein was reputedly sweating while shoveling snow, a temperature of 10 degrees below zero Fahrenheit and a wind of 30 mph would produce a wind chill in the 65 below zero range.[26] Under the new values, which went into effect in November 2001, to produce a similar wind chill with a 30-mph wind would require a temperature of 30 degrees below zero.[27] Gee, no wonder Bluestein was sweating on that January night in Indianapolis.

For its numerous, acknowledged imperfections and incompleteness, the new wind-chill index represented a significant upgrade over the Siple–Passel scale.

FREEZING, AND NAKED SUNBATHING

Paul Siple, who accompanied Admiral Richard Byrd on all five of his expeditions to Antarctica, would become something of a national celebrity and once graced the cover of *Time* magazine. He made his first trip to the bottom of the earth in 1928, chosen as an Eagle Scout to join Byrd, and made an appearance in the 1930 documentary, *With Byrd at the South Pole*. Of note, the Antarctic coast has several geographic features named for him, include Siple Island and Mount Siple.[28]

For his part, Charles Passel joined a Byrd mission that began in November 1939, with stated goals including mapping the coastline and searching for minerals. Passel had a wide variety of duties—supervising supplies, sled-driving, caring for the sled dogs, operating the radio.[29]

According to his published 400-page diary, Passel was stationed at the West Base, where he was the geologist and Siple was the leader of the scientific staff.[30] Their most famous collaboration led to the introduction of one of the most popular catchphrases ever to enter the meteorological lexicon. It involved nine ounces of freshly melted snow poured into a narrow plastic container. The container was attached to the top of a tall pole, and wires transmitted weather data to the science building every five minutes.[31] Thus, they were able to measure the heat that was lost by the combination of wind and cold.[32]

What did that have to do with winds blowing in our faces on a January morning?

Siple and Passel observed that the stronger the wind, the faster the water became ice. When water turns to ice, it surrenders heat to the colder air, and wind speed accelerates loss.[33] Taking that to the levels of flesh and blood, our bodies surrender heat to the cold air as winds blow against us. The stronger the wind, the more rapid the loss, and the colder we feel. Blow on a hot cup of coffee, and it will lose heat. The atmosphere blows on our face, we lose heat.[34]

In the groundbreaking paper they published in April 1945, Siple and Passel introduced the term "wind chill," although the version that the National Weather Service eventually shared with the public was somewhat different from the original iteration.

Siple and Passel developed a complicated scale itemizing the "stages of relative human comfort" in a range of 0 to 2600. At level 100, for example, they said "nude sun-bathing" is "possible"—yes, even in Antarctica—"but eyes must be protected." (Parental discretion suggested.) At 2300, "Exposed areas of face will freeze within less than a half minute for the average individual."[35]

Meteorologists eventually figured out how to derive a scale that would be more palatable to the public, and the National Weather Service began using wind chill temperatures in 1973.[36] When you hear something on the order of "The winds will make it feel like it's 5 degrees below zero," that's more likely to get your attention than, "The winds will make it feel like a 1400 on the Siple–Passel index."

Like that index, the Osczevski–Bluestein scale has its flaws, but has become the twenty-first-century standard for both the U.S. National Weather Service and Environment Canada. On another level, it provides the basis for a realistic assessment of frostbite dangers.

A DIFFERENT FEEL

Not that Siple and Passel failed to develop metrics for the frostbite risk. They did. And they may have been perfectly useful—if you lived or happened to be vacationing in Antarctica or the North Pole. The frostbite-hazard on their scale began at 27 degrees below zero Fahrenheit and bottomed out at 148 below. Not a whole lot of people live in places that experience those extremes.

Working with the twenty-first century wind-chill values, Michel B. Ducharme and Dragan Brajkovic, two Canadian defense department researchers, undertook a major project to develop frostbite thresholds that would be applicable where most people lived, and would be based on heat loss on the human face, rather than on a plastic bottle. To do this right, they needed 12 actual human faces. As it turned out, six men and six women who met certain body type and height requirements did volunteer.[37]

In the summer before the wind chill index became the U.S.–Canadian standard, the volunteers reported to the Civil Institute of Environmental Medicine in Toronto, where they had the pleasure of being placed in a "chilled" wind tunnel. Instruments measured the heat flow from their faces as they walked 3 miles per hour on treadmills.[38] The temperatures to which they were subjected ranged from 32 degrees Fahrenheit to 50 degrees below zero. While they walked, the wind speeds were varied from 10 to 20 mph, but mercifully kept to 0 mph when it was 50 below zero.

This was not a sadistic exercise. Their entire faces were exposed, but the subjects wore "wool socks, mukluks and heavy mitts during all exposures and a ski hat was worn which covered the ears," Ducharme and Brajkovic wrote. And the volunteers were monitored closely for "frostnip."[39] Frostnip causes no permanent damage to the skin, but it is the first stage of frostbite.[40]

Each subject participated in six scheduled 45-minute sessions over a three- to five-day period; however, the sessions were halted with the first signs of frostnip, in which ice crystals form on the top layers of skin. The "white or yellowish firm plaque on the skin is easily detectable," the researchers said. "The frostnip was detected very quickly by an investigator inside the cold chamber and the skin was rewarmed within 30 seconds. . . . The investigator placed his warm palm on the frost-nipped site until a normal skin color returned."[41] That might be a source of handy advice the next time you experience frostnip and you have the good fortune to be with someone who has an available warm hand.

Thanks to the volunteers, the public now has realistic thresholds for frostbite that are useful outside the polar regions. For example, we know that if outside for a half hour when the temperature is zero degrees Fahrenheit and the winds are blowing at 15 miles per hour, the wind chill is 19 below zero and chances are good that your skin could freeze within a half-hour of exposure. Of course, that threshold will vary depending on whose skin we are talking about and what that person is doing outside, other than perhaps showing questionable sense.

Osczevski and Bluestein said the day may come when tailored charts will be available for, say, skiers, or will take into consideration what difference it makes when damp winds blow off the ocean. For now, they wrote, the public appears to be satisfied with what they called "a deceptive simplification."[42]

While the Siple and Passel wind-chill calculations had serious deficiencies, neither Osczevski nor Bluestein, nor Ducharme nor Brajkovic, nor anyone else who has made a serious attempt to answer the "how does it feel" question in the past—and maybe into the future—would take issue with a statement that Siple and Passel included in their 1945 paper: "So many variables are involved that is unlikely any scale can ever be employed which will satisfy all conditions."[43]

Yes, as Osczevski recounted, it may well be true that a sweater is something you wear when your mother is cold. But if you're both outside and wind chill is −19 degrees Fahrenheit, you'll both need sweaters, and if you're shoveling or doing something strenuous, the National Weather Service advises that you're better off with synthetic fabrics. Stay away from cotton. Once it gets wet, it takes time to dry, and wetness will drain your heat.

Here are other dress-for-the-cold tips from the National Weather Service.

- Wear tightly woven, water-repellent, hooded outer garments.
- Cover the mouth to spare the lungs from extreme cold.
- Mittens, snug at the wrists, are better than gloves.
- Top it all with a hat: About 40 percent of body heat escapes from the chimney of the head.[44]

CHAPTER 11

THE BRIGHT SIDE OF COLD

- The cold has a long history of therapeutic uses, and the people who participate in those "polar bear plunges" may be on to something.
- Snow shoveling can be hazardous, but exercising in the cold can benefit the heart and help the body produce healthy "brown fat" and burn excess calories.
- Winter is a friend to pollen-allergy sufferers, and cold can save you from the world's most dangerous animal.

+ + +

He wasn't wearing layers, synthetic or otherwise. No scarf. No mittens. He assuredly wasn't abiding by any wind-chill or frostbite charts. He was bootless, shoeless, sockless.

Wim "the Iceman" Hof was running in shorts. Barefoot. All the more notable because he was running a half-marathon, 13.1 miles. All the more notable because he was doing this in the Lapland region of Finland. On the poleward side of the Arctic Circle. In January.[1]

Hof has gained international celebrity (and a measure of infamy) as an ambassador of the cold, proclaiming the message of exposure to, and immersion in, the cold and "how good this is for our health and mentality."[2] Along the way, he also has gained considerable notoriety, not to mention legal headaches and accusations of endangering lives, even to the point of death.

Yet Hof may possess one of the world's more studied bodies, and some health experts, even skeptics, do agree that he is a physical marvel. One 2018 study concluded that he was capable of allowing his autonomic nervous system to adapt to the cold.[3] That's the critical part of us that regulates involuntary responses, such as heart rate and blood pressure.[4] And medical researchers have found at least some of what he says about our bodies and the cold to be true.

I was introduced to Hof's exploits by our younger son, an avid winter hiker who did the snowy Mount Mansfield summit six times in a week the last time we were up that way and who skied there on a day when the wind chill was 60 degrees below zero. As much as I love winter, that's a tad colder than I'd prefer to be.

Then I started reading about Wim Hof. By comparison, what our son did was on the order of taking walks in Central Park in April. I found Wim Hof to be a bit other-planetary for my tastes. As of this writing, I haven't changed my mind.

For his efforts in Lapland, which he undertook at the behest of the Discovery Channel, Hof developed a "devastating problem" that lingered for weeks. During the race, he felt a disconcerting sensation in his left foot. "My whole foot resembled that of a wooden stick," he recalled in an autobiography. He had suffered third-degree frostbite on his feet and returned to his native Holland on crutches.[5]

Medical experts, not surprisingly, have varying perspectives—some of them quite unflattering—on the unconventional wisdom of the Iceman. They do agree that whatever he is proselytizing about, it certainly isn't for everyone. They also concur with him on one significant point: Cold can be good for you. In fact, it has any number of direct and indirect positive impacts.

Cold often has been exploited for its therapeutic value. As the Harvard Medical School researchers have pointed out, people in Finland and Russia have viewed cold-water immersion as being beneficial to health. In one Finnish study, for three months 10 women periodically immersed themselves in nearly ice-cold water for periods of 20 seconds, and blood tests showed a jump in levels of a pain-suppressing norepinephrine.[6] So-called cryotherapy—the use of very cold air, ice, or cold water—has been used for centuries for pain relief for sports and other injuries.[7] As the Harvard team observed, "a little bit of exposure may not be such a bad thing." Others would even insist on making it more than "a little."

Ever wonder why cold packs are recommended so often to quell swelling and pain? Yes, the evidence again and again has thrown considerable quantities of cold water on the notion that cold is all bad.

BENEFITS OF "COLD WATER IMMERSION"
"Polar Bear Plunges," in which normally sane people don bathing suits for the pleasure of wading into bodies of water in the planet's colder regions, have become wildly popular fund-raising events. They have been monster rain-raisers for the Special Olympics over the years. Thousands upon thousands of people participate, more than 12,000 one year at the Maryland plunge into the Chesapeake Bay.[8]

I have no regrets about my lack of participation in these clambakes, but perhaps I should reconsider. Those Finns and Russians evidently are on to something. These encounters with refrigerated waters do appear to hold some health benefits.

The concept of ice and cold-water bathing, properly called "cold water immersion," or CWI, has been gaining traction around the world in recent years. In a review of more than 100 studies, Norwegian researchers noted that "many" of those studies documented that CWI had "significant effects." They acknowledged that the studies had their limits, but the potential CWI benefits were unmistakable. For example, CWI "may have a protective effect against cardiovascular, obesity, and other metabolic diseases" by helping to reduce "adipose tissue," according to a paper in the *International Journal of Circumpolar Health*.[9] Colloquially, "adipose tissue" is known as body fat.[10]

University of Oregon physiologist Dr. Chris Monson, quoted on the Healthline website, said that frequent cold-water exposure may help reduce inflammation. He also pointed to evidence that it may be a boon to blood-sugar regulation. That's a positive side effect that could lower the risk of diabetes and be good for the heart.[11]

Of course, before you experiment with CWI, consult with your doctor.

COLD AND THE HEART

How often are we warned about the perils of snow-shoveling—as well we should. A 17-year comprehensive study documented that on average, shoveling snow results in more than 11,500 injuries annually, including slip-and-fall and lower-back cases. In addition, it is blamed for about 100 deaths every year in the United States.[12]

It can be especially dangerous for people with underlying heart conditions.[13]

Yet for the healthy, snow-shoveling "can be a great way to fit in some moderate exercise," writes Dr. Aaron Lee, a sports-medicine specialist with Loyola Medicine. Make sure to warm up before shoveling, maybe taking a short walk. Push, rather than lift, the snow as much as possible.[14]

While cold ratchets up the stress on the heart, exercising in the cold also can benefit heart health.[15]

In lower temperatures, the heart works harder to distribute blood throughout the body. Working out in the cold can strengthen the body's circulatory system and lungs.[16]

Just be smart about it: Snow-shoveling is not an Olympic competition. And if you have any trepidation, consult with your physician.

THE BROWNING OF THE FAT

Cold stimulates the production of so-called "brown fat," a somewhat unflattering term for something that evidently we could all use more of.[17] It is the good—and scarcer—variant of that "adipose tissue." The more-familiar "white fat" dominates in our bodies, and even that's not all bad: White fat does protect our organs. Unfortunately, too much of it, especially in the belly and thighs, is a symptom of obesity. We are seeing the results of too much of it all over the country.

Brown fat breaks down blood sugar while it produces heat.[18] The color is the result of the fact that it contains rich supplies of mito-chondria, which are storehouses of iron and generate the energy that is required to power our cells.[19] And research has shown that exercising in the cold actually can transform some of that white fat into brown fat.[20] Plus, the body keeps using energy hours after a workout in the cold, resulting in calorie afterburn.[21]

NOTHING TO SNEEZE AT . . .

Of all the positive aspects of the cold, a personal favorite it that it is death to allergenic pollen. As someone who has had severe allergies to tree, grass, and ragweed pollen and airborne mold spores, I can't thank winter enough. Cold kills the reproductive moods among the plant and fungal life.

I am well aware of the indoor allergenic hazards. I can count on a solid round of sneezing when the heating systems have their first runs in the fall. But a day or two of congestion beats the hay out of six months of it. When the pollen counters stop counting around the time of the autumnal equinox, for millions of us, that's a champagne-popping situation.

THE BITE ON THE BITE

What is the world's deadliest animal? The shark, lion, tiger, cobra snake? None of the above, and it turns out that the cold can save us from an attack by the animal the Centers for Disease Control has crowned as the "deadliest."

The mosquito.

It is hard to find enough pejoratives to place at the tarsus of the mosquito. They would be just plain annoying if they weren't so dangerous. An outbreak of West Nile virus, the number one mosquito-borne threat in the United States, sickened more than 1,700 people in Maricopa County, Arizona, in 2021. That same year, mosquito-borne malaria was blamed for killing more than 600,000 people in 84 countries.[22]

The air may have an unpleasant bite on a frigid morning, but be thankful that you're in no danger of a mosquito bite, or a tick bite, or most any other kind of potentially sickening insect encounter when you're outside.

Mosquitoes are around, biding their time until the weather gets warmer, and on occasion you might even see one on a balmy February day even in places where it's been known to snow in winter.[23] But if it's below freezing or thereabouts, you'll neither see nor hear a mosquito, wearing layers or otherwise.

CABIN FEVER SURE CURE

One of the perhaps underrated impacts of the COVID-19 outbreak was the epidemic of social isolation during the lengthy lockdowns. What resulted was an epidemic of "cabin fever,"[24] another of those diagnoses for which you will find no antidote in the *Physicians' Desk Reference.*

It is loosely defined, but I'm partial to the description by Paul Crawford, director of Britain's Centre for Social Futures at the Institute of Mental Health, who described it as "commonly understood to refer to a combination of anxiety, lassitude, irritability, moodiness, boredom, depression, or feeling of dissatisfaction in response to confinement."[25]

Cabin fever is often associated with snow and cold, and the term often is invoked in winter. Mental- and physical-health experts offer any numbers of tips to combat cabin fever, among them the most obvious therapeutic strategy: Get outside.

For one thing, your body should appreciate a natural dose of vitamin D. The sun's daily wattage really accelerates in February over the places that experience real winter. In Chicago, for example, the wattage at the end of February is about the same as it is around Columbus Day in October.[26] Vitamin D supplements can be helpful, but why not consult the primary natural source if that's at all possible?

If you can avoid cabin fever, by all means do so. And if you wake up one morning and the wind-chill index looks particularly forbidding, you might consider the 350,000 hardy folks who live in Yakutsk, in the Yakutia province of Siberia.

"LIKE A CABBAGE"

Nurgusun Starostina works at a market where she sells frozen fish. One of the advantages of selling frozen fish in Yakutsk is that you don't need a freezer. You're living in one.

On a January week in 2023, the temperature got down to 58 degrees below zero Fahrenheit. Yakutsk is believed to be the coldest city on earth. Certainly, it is the coldest city on earth with a population of 350,000.

Starostina doesn't see anything particularly heroic about withstanding temperatures that would be unimaginable in any well-populated area of the United States, including Alaska. "Just dress warmly," she told Reuters. "In layers, like a cabbage!"[27]

Urban researcher Ksenia Acquaviva, formerly with the Skolkovo Center for Urban Studies in Moscow and who was involved in an international research project based at George Washington University, has described Yakutsk as "impressively vibrant and rich in cultural and urban life." She wrote in 2017 that after experiencing stunning population growth, it was "still actively attracting people from neighboring regions."[28] Dun & Bradstreet lists over 1,500 manufacturing companies in the city.[29]

The city's ice-cold reputation actually has lured tourists, and among its attractions is the Mammoth Museum, where woolly mammoth fossils are on display.

Yakutsk is not without quality-of-life issues. It has its share of slums and has struggled with budget deficits. And, ironically, worldwide warming may be a threat to its future, said Acquaviva.[30] The town is situated atop permafrost, defined as ground that remains at or below freezing for two or more straight years. Maintaining that state of permafrost is critically important for stabilizing building foundations.[31] The city "has a long-standing experience of design and construction in extreme weather conditions," Acquaviva observed, but thawing permafrost is causing visible warping among buildings, road surfaces, and pavements.[32]

The Siberian regions have gained international attention recently for record warmth, but it is still plenty cold there, and the resiliency and hardiness of the population is remarkable by any standard.

Yakutsk evidently has a historic distinction in the annals of life in the cold. It involves one Semennikova (Djakonova) Varvara Konstantinovna, who is believed to have been the longest-lived person on the planet.[33] She died in 2008, two months shy of her 118th birthday.[34] Her age was documented through the records of a Yakutsk church. Researchers who investigated her life observed in a paper published in the journal *Advances in Gerontology* that she had "not suffered from serious diseases." They concluded that her longevity was "the unique example of the extremely high level of adaptation to the extreme climate."[35]

No evidence suggests that Semennikova's case spoke to a legacy of longer life spans in her province, although they did increase significantly from 2021 to 2022, from about 70 to 73, according to CEIC, the global economic database.[36] Her remarkably long life and the growing population of Yakutsk under such harsh conditions arguably speak to the extraordinary adaptive powers of our species.

ABOUT WIM HOF

"Most people just think, who is that crazy man?" Judging from the vast amount of written and visual materials about Wim Hof's lifestyle and practices, that observation appears to be a reasonable self-assessment. "But we've got to get back to the cold," Hof told author Susan Casey, who wrote a flattering and wonderfully detailed profile of him for *Outside* magazine. And with that comment, Wim Hof walked into a glacially cooled lagoon in Iceland. At one point she described him being in a stare-down with a seal that he greets with a "Hi Johnny!"

Who is that crazy man? Hof emerged a half-hour later, skin lobster-red, but not a trace of a shiver. Susan Casey later would join a group that spent two minutes submerged in ice water and declared it so exhilarating, she would continue the practice when she got home.[37]

Professor Wouter van Marken Lichtenbelt, a Dutch physiologist who has spent time with Hof, is among those who suggest that the Iceman's concepts may indeed hold some value. "It doesn't hurt to try," he wrote, but he also raised cautionary flags. Wim Hof's claims still need scientific validation, even if "people may feel healthier."[38]

British Professor Mike Tipton, of the University of Portsmouth's Extreme Environments Laboratory, warned that some of Hof's practices clearly would be dangerous for a whole lot of people. "For example, standing in snow barefoot for 30 minutes introduces the risk of cold injury," he says. "Plunging into cold water introduces the risks of drowning and cardiovascular problems."[39]

OUR MINDS AND THE COLD

In the end, our key to surviving the cold has everything to do with something that we would always keep under our hats when cold—our brains. So say the anthropologists. Our brains are what really got us this far and led human beings to create habitats from the Equator to the polar regions.

A Wim Hof notwithstanding, our species did not have the advantages of the biological adaptability of other primates. But what we have lacked appears to have worked to our benefit. This deficiency, a team of researchers observed in *Scientific Reports*, is likely "due to the importance of human behavioral adaptations, which was a buffer from climatic stress and were likely key to our evolutionary success."[40]

One big advantage that "behavioral adaptation" has over "physical climatic adaptation" is that it's a heckuva lot quicker. "Humans are the ultimate adapters, thriving in nearly every possible ecological niche."[41]

Humans have developed excellent clothing, houses, and behavioral adaptations to cold, and these seem to be more important for those living under extreme conditions than our physiological mechanisms alone.[42]

That ingenuity and adaptability has put us in positions to live in wintry environments and to mine the cold for the health benefits it can offer. We didn't do so well on the fur-growing front as a species, but through the eons we have learned to protect ourselves from the worst of cold's hazards.

While the benefits of some of the extreme activities in cold environments that are championed by Wim Hof are questionable and need vigorous investigation,[43] health experts assure us that spending some time in the cold can be good for us mentally and physically. And you don't have to plunge into ice water, or stare down seals.

Here are some practical tips for exercising safely in the cold, from Everyday Health.[44]

- **Warm to the core.** Above all stay dry. Start with a thin synthetic layer. Add polar fleece, if it's especially cold, and a shell. You can peel off layers if you get sweaty.

- **Watch your steps.** Running shoes are built to allow heat to escape. In the cold, consider shoe covers, which are sold in winter-sport shops. Also, when hiking or running on snow or ice, I highly recommend MICROspikes for traction. They really work.

- **Moisturize, internally and externally.** In addition to drinking liquids even if you're not thirsty, keep your skin moisturized. Cold air holds less water vapor than warm air, and that dryness takes tolls on the skin.

CHAPTER 12

VITAMIN D'S DAY IN THE SUN

- Credible research has made cases for the benefits of "the sunshine vitamin," including its effects on bones, heart conditions, cancer, influenza, and overall well-being.
- The sun remains the primary natural source of vitamin D, but it also is available in some foods—even Cocoa Puffs—and supplements.
- It is estimated that a billion people worldwide have vitamin D deficiencies, but precisely how much the body needs remains "a source of never-ending debate."

+ + +

Through the years, we have witnessed so many nutritional fads come and go and come back again. Vitamins B, in its various forms; C; and E all have had periods of celebrity in the last 50 years—and unquestionably with some justification. These days, perhaps related to the COVID-19 pandemic and its aftermath, vitamin D, "the sunshine vitamin," literally is having its day in the sun.

Testifying to the consensus notion of its paramount importance is the fact that cereal giant General Mills announced in the summer of 2023 that it was preparing to fortify some of its popular brands with enough vitamin D to supply 20 percent of the daily minimum requirement with each bowlful. The cereals named weren't exactly the favorites of the farm-to-table, whole-grain/organic crowd: The list included

Honey Nut Cheerios, Lucky Charms, Cocoa Puffs, Trix, and Cookie Crisp. With "96% of all Americans ages 2 years and older falling short on this key nutrient," this clearly was an epidemic situation, according to the company's news release.[1]

While the sourcing on that 96 percent figure was quite unclear—and for the record the National Institutes of Health has put that figure at 25 percent[2]—the message was clear: If General Mills was joining the war, vitamin D must be a matter of survival.

But it is reasonable to ask: If vitamin D indeed is the sunshine vitamin, then why should I need to eat a bowl of Cocoa Puffs to get it?

VITAMIN D'S PRIMARY SOURCE

It really is the sunshine vitamin, and you can get it even if you have a Cocoa Puffs or Honey Nut Cheerios intolerance.

Vitamin D has gained ever-more prominence for the simple reason that it is in fact essential. It is estimated that as many as a billion people worldwide don't have enough of it in their bodies.[3] Deficiencies have been linked to maladies from poor bone health, to cardiovascular disease, to influenza, to overall immunity issues.

Mountains and mountains of papers, some quite excellent and provocative, others mildly inscrutable, on the subject have appeared in respected journals, and as is so often the case in matters of biometeorology, studies have shown most everything. A certain perplexity among lay readers—even among members of the medical profession—would be understandable.

What is evident is that we are blessed with an indefatigable natural source, a gift from the solar system. Underscoring its importance, its presence is evident in the healthy, even it may take a few more lifetimes to unravel the cause–effect relationships. Dr. Michael Holick, the pioneering—and controversial—researcher, has gone so far as to say that if people had what he calls "optimal" levels of vitamin D, health-care costs would drop by 25 percent.[4]

In winter, it can be a challenge to capture the sunshine vitamin in its natural state, especially for those who live in the higher latitudes—say, north of the Mason–Dixon Line in the United States—but some health experts advise that it may be a challenge worth accepting. The sun does shine on some winter days, even in places like Syracuse, New York, and our bodies don't ask a whole lot of us.

It would take maybe two hours, but even on a cold winter day in Boston a body that was 95 percent covered still could produce a daily supply of vitamin D, estimated Dr. Robert Ashley at the University of California, Los Angeles.[5] Researchers at Tufts Medicine, in not always sunny Boston, opined, "Every bit helps. If possible, try to get outside each day to take a walk. Just 20 minutes of sun several times a week can make a difference."[6]

If you would like a second opinion, consult your doctor. If you would like third and fourth opinions, consult more medical professionals.

A VERY BRIEF HISTORY

Vitamin D's celebrity is a modern phenomenon, yet the first known link between sunshine and bone health dates back about 2,500 years. It is credited to the Greek historian Herodotus. While visiting the battle-field where King Cambyses prevailed over the Egyptians, he examined the skulls of the slain combatants. He found a puzzling phenomenon. The Persians had skulls so fragile that they broke when hit with a mere pebble, whereas the Egyptians' skulls clearly were far stronger. Egyptians with whom he spoke explained to him that from childhood their heads were bare, whereas the Persians wore turbans. Could the sun have that much of an impact on the strength of bones? (Evidently, however, it took more than harder heads to prevail in battle. The Persians won.)[7]

More than two millennia later, Cambridge University physics pro-fessor Francis Glisson made another, seminal bone–sun connection. He observed that cases of debilitating rickets were common among children of farmers. Even though they "ate well," they lived in cloudy, rainy parts of the country and were kept inside during the winters.[8] Efforts to erad-icate rickets, a softening of the bones that can result in delays in growth and motor skills, severe spinal pain, muscle weakness, and skeletal defor-mities,[9] would be critical to linking vitamin D with bone health.

It was the lengthy search to find a cure for rickets that led to the identification of vitamin D in 1920. It wasn't long thereafter that the movement to fortify foods with vitamin D was underway, and rickets became rare.

THE VITAMIN D FACTORY

While it was one of 13 vitamins discovered in the early twentieth century as doctors were researching diseases caused by nutritional deficiencies, vitamin D is unique in that it is produced by the human body. Even when it is consumed in foods, it must undergo a transformation.[10]

And its natural presence in foods is limited. Among the best natural sources are fatty fish and fish-liver oils, with smaller amounts in egg yolks, cheese, and beef liver. It is also available in larger quantities in supplements.

But the primary source of natural vitamin D was and still is the sun.[11] "Sunlight is unarguably the most optimal way to obtain vitamin D," wrote Sneha Baxi Srivastava, pharmacy-education specialist at Rosalind Franklin University of Medicine and Science in the sometimes sunshine-deprived Chicago area. Besides, she noted, "It is free," and while the sun's ultraviolet B (UVB) rays may be a source of concern for the skin, they never would be a source of vitamin D "toxicity" because "our body is able to self-regulate the amount of vitamin D it receives."[12]

How we produce vitamin D from sunlight is somewhat complicated, but here is a very simplified version: When the sun strikes the skin, our body makes vitamin D from cholesterol,[13] which, despite the reputation the very term evokes, isn't all bad. Harvard Health experts describe it as a rather physically unattractive "waxy, whitish-yellow fat," but add that it is "a crucial building block in cell membranes."[14] The body then employs a two-step process involving the liver and the kidney to manufacture active vitamin D.[15]

The sequence first was elucidated by Dr. Holick in a 1980 study in which he and his colleagues experimented on rodent skin.[16] They described a "unique mechanism for the synthesis; storage; and slow, steady release" of vitamin D from the skin into the circulatory system.[17]

CELEBRITY DOCTOR

For his work, Dr. Holick developed a certain celebrity and helped raise the profile of vitamin D. In a 2010 interview with *Life Extension* magazine (vetted in 2023), Holick recalled that at one time his proselytizing about vitamin D had elicited an epidemic of yawns. He evoked the image of a famous commercial figure whose services were unneeded. "I felt like a Maytag repairman every time I got up to talk to doctors about vitamin D," he recalled.[18]

While the ties between vitamin D and childhood bone health had been well-established by the early twenty-first century, Holick, as much as anyone, was responsible for popularizing the linkages between vitamin D and a whole menu of health issues.

Holick has gained his share of detractors. He evidently became quite wealthy by extolling the virtues of supplements, receiving hundreds of thousands of dollars from the vitamin D industry, including drug companies and indoor tanning concerns, according to KFF Health News.[19] But health experts do not dispute vitamin D's importance in our bodies' hierarchies of needs.

NO BONES ABOUT IT

In tandem with calcium, vitamin D is a pillar of bone health. Just about all of it is in our bones and teeth.[20] Having low bone density can be dangerous because this makes your bones more susceptible to breaking, even without traumatic injury.[21]

Our bodies can't manufacture calcium, so we need to get it from the things we consume, and vitamin D plays the critical role of helping us take in calcium, which is essential for our nerves, muscles, and hearts to function properly.[22]

THE HEART OF THE MATTER

Although the physical linkages aren't always clear—and research has argued against the value of supplements in cardiac matters—not in dispute is the correlation between healthy levels of vitamin D in adults and healthy hearts. It is known that vitamin D is a regulator of blood pressure, and deficiencies have been tied to cardiovascular "risk factors."[23]

In the *Life Extension* interview, Holick stated categorically that "optimal levels" of vitamin D would reduce the incidence of heart disease.[24] However, his dosage recommendations appear to be 40 percent or more higher than the estimates in a 2010 study.[25]

"There is some promising research that vitamin D may reduce the risk of heart failure, and that requires additional research," said Dr. JoAnne Manson, the lead author of the expansive VITAL study—VITAL stands for Vitamin D and Omega-3 Trial—that involved more than 25,000 participants age 50 and older. Manson is a professor of medicine at Harvard Medical School and chief of preventive medicine at Brigham and Women's Hospital in Boston.[26] According to Manson, "multiple factors" may explain the relationship between heart health and higher vitamin D levels. While people with those higher levels have been found to be "less likely to have cardiovascular disease," separating the specific effects of vitamin D versus those other contributors would be problematic, she said in an article published on the American Heart Association website. Take exercise, for example. "People who spend more time outdoors engaged in physical activity, which supports heart and vascular health, may have higher vitamin D levels from incidental sun exposure," Manson reasoned.

Diet would be another consideration. A number of foods higher in vitamin D, including certain fish, tend to be heart-friendly.[27] Those choices also would fit the profile of foods consumed by people with healthy eating habits.

They also are less likely to be obese.

OBESITY

Obesity has been coupled with any number of ailments, and medical researchers have documented clear correlations between vitamin D deficiencies and obesity.[28]

Could it be a volumetric question? One possible explanation for the lower levels of vitamin D among obese people is that it is stored in fatty tissues. It makes sense that those with an excess amount of body fat need to acquire higher amounts of the vitamin just to reach levels that are considered healthy.[29]

Holick held that deficiencies contribute to obesity because weak bones and muscles are serious impediments to movement and physical activity. Even without eating more, people who don't move are going to gain weight.[30]

Cause-and-effect relationships, however, remain a subject of research and debate. "Low vitamin D levels can't be ruled out as a cause of obesity," said Caroline Apovian, obesity and nutrition specialist at Brigham and Women's Hospital and Harvard Medical School.[31]

CANCER

The debates and controversies over the value of vitamin D supplements appear to have an infinite shelf life, the VITAL findings notwithstanding.[32] However, at the very least correlations between deficiency and cancer have been identified, if not the causes. Research has established that people who live in places where the sun shines longer and more powerfully tend to have lower rates of cancer. The matches have been imperfect, but some studies link vitamin D with lesser chances of being stricken with colorectal cancer.[33] The ties to bladder cancer haven't been as strong, with virtually no associations found with breast or lung cancers.[34]

While research has documented that breast cancer patients had insufficient vitamin D levels at the point of diagnosis, it is unclear whether this was the cause or consequence of the cancer.[35]

In any event, Holick said it wouldn't be surprising to find deficiencies in cancer patients under treatment. They have to be wary of sun exposure because of the sensitivities attendant to chemotherapy. They also suffer so much from nausea that it's likely they would not be eating well-balanced meals. The deficiency compounds their health problems by contributing to muscle weakness and aches and pains.[36] Again, Holick has been one of the nation's most steadfast advocates for supplements.

That VITAL study did find a reduction in the risk of cancer death among those with normal body weight who used supplements.[37] A total of 793 cancers occurred among the 12,927 supplement-using participants, as compared to 824 among the 12,944 in the placebo group. That would be about a 4 percent difference. And, again, weight clearly made a weighty difference. Among 7,800 participants classified as having "normal" weight, 58 of those in the vitamin D group developed metastatic or fatal cancers. In the placebo group, the total was 96. However, the researchers found no evidence that supplements could prevent the onset of cancers. The cancer-onset risks were about the same for both the supplement and placebo group. The difference was that the supplement group's risks of metastatic or fatal cancers was lower.

Manson said in an article posted in 2023 that studies on animals suggested vitamin D may have some effects on how tumors develop, perhaps making them less likely to spread. "We're not recommending that everyone routinely take high-dose vitamin D for the prevention of advanced cancer," the article quoted her as saying. That said, those with high cancer risk should be sure to get at least the minimum vitamin D dosage daily.[38]

As for which cancers were most affected by the supplements, researchers were unable to drill down on conclusions about specific types.[39]

INFLUENZA . . . AND SUNSPOTS?

Influenza outbreaks typically peak in the winter months when the natural source of vitamin D is wanting in the temperate zones. Is that more than coincidence? The answer is a resounding yes, according to some researchers.

Dr. Robert Edgar Hope-Simpson—the British general practitioner and renowned medical researcher who proved through a painstaking investigation that chickenpox and shingles were caused by the same virus[40]—connected the dots between the annual solar rhythms and the influenza incidence in a paper published in 1980.

Hope-Simpson had gone so far as to postulate that the solar storms that we commonly call "sunspots," which electrify the night skies with displays of the aurora borealis, were related to flu epidemics. He held that peaks in sunspot activity in 1937 and 1828 coincided with outbreaks and that the 1917 maximum "anticipated" the deadly Spanish-flu epidemic that followed. He acknowledged, however, that flu records probably were "too dubious" to make a thorough investigation of the possible linkages worth the effort.[41]

Hope-Simpson did make a case for solar linkages with flu outbreaks by observing that epidemics tended "to occur contemporaneously at the same latitude even in localities widely separated by longitude. The explanation must be that the epidemic process is approximately synchronous." The epidemics "often occur in different months, but the tendency to latitudinal synchrony is evident." In the temperate latitudes in both the Northern and Southern Hemispheres, the preponderance of influenza cases occurred in winter.

Hope-Simpson suggested that diseases have something in common with the foliage and the annual rhythms of birds or mammals. "Seasonal disease in man is no exception to the law that all seasonal phenomena are caused by variations in solar radiation," he said.[42]

Other researchers, including Holick, have picked up on Hope-Simpson's hypothesis of a sunspot–influenza connection. Holick was among the authors of a paper published in 2006 that took Hope-Simpson's findings to another level.[43] They noted that a radiational explanation for respiratory illnesses in winter is that people spend more time congregating indoors, thus increasing the potential for the contagion.

However, what would explain why outbreaks are infrequent in summer, "in spite of people congregating on cruise ships, airplanes, nursing homes, factories, etc."? As Hope-Simpson documented, flu was primarily a wintertime ailment. He clearly had identified a "seasonal stimulus," but lacked an explanatory mechanism. They were confident they had found it, postulating that because solar radiation is a robust source for stimulating vitamin D production, which has "profound effects" on immunity, it would make sense that in winter months, vitamin D deficiencies would be common.

Vitamin D, they wrote, was "a likely candidate" to explain Hope-Simpson's "seasonal stimulus." The researchers concluded unambiguously in 2006 "that sunlight strongly protects against getting influenza."[44]

Anyone who has had a nasty bout of influenza is aware that it can be physically debilitating, with the aftereffects lasting for weeks. I recall having a case of it so bad that it left me in a state of hallucinatory weakness.

Besides the physical effects, influenza annually has tremendous impacts on workplace productivity, and economists David J. G. Slusky from the University of Kansas and Richard J. Zeckhauser from Harvard Kennedy School conducted an econometric-statistical analysis to assess what sunshine and vitamin D have to do with it all. They suggested that our walks in the sun can benefit our fellow workers and the U.S. economy. "This study reinforces the long-held assertion that vitamin D protects against acute upper respiratory infections," they wrote in their 2020 paper. "One can secure vitamin D through supplements, or through a walk outdoors, particularly on a day when the sun shines brightly. When most walk, through herd protection, all benefit."[45]

No question, "Vitamin D is essential for optimal health outcomes," observed Sneha Baxi Srivastava, the Rosalind Franklin pharmacy specialist. "Although the evidence may not be conclusive about its impact . . . the significance of maintaining adequate vitamin D levels is indisputable. . . . As is the data presenting how humans are spending much less time outdoors."[46]

HOW MUCH IS ENOUGH?

I cannot speak for our readers, but as a lay vitamin D consumer, I found the pursuit of that question to be a dizzying experience.

The amount of vitamin D we get from sunshine depends on several variables—including the time of day, the extent of cloudiness, the UV index, where you are on the planet, and the darkness of your skin. Contrary to what the Tufts' people and UCLA's Dr. Ashley suggested, a 2013 paper coauthored by Holick stated that the sun over Boston would provide no vitamin D from November through February. (It was unclear if he was including leap years.)[47]

The National Institutes of Health do provide metrics for daily consumption in micrograms—or mcg, equal to one-millionth of a gram—and international units—or IU, a measure of activity. For most people the recommendations are 15 mcgs and 600 IUs, but 20 mcgs and 800 IUs for those ages 71 and over.[48] For those keeping score, that's about four bowls of Cocoa Puffs for most folks—better make it five for those 71 and over.

Holick has insisted that we need considerably more than NIH suggests.[49]

Don't be too surprised if those guidelines shift in the years to come, and outsized—and undersized—recommendations pop up in research findings. To quote Dr. Caroline Apovian, the issue of how much is best for the body "has been a controversial topic of never-ending debate in the medical literature."[50]

In the meantime, say Yale Medicine doctors, the nation is not confronting a vitamin D "epidemic," General Mills notwithstanding. "Most people should be fine," according to Dr. Karl Insogna, director of Yale Medicine's Bone Center.

Testing of vitamin D levels is recommended for those whose skin has limited exposure to the sun and who avoid dairy products in their diets. Insogna also suggested at least one test for those over 70 years of age. Providing they are not overdone, supplements are perfectly fine, especially for those with lactose intolerance and milk allergies, fat absorption issues, and for people with darker skin.

Here is additional guidance from health experts:

- **Diet.** As a vitamin D supplier, the sun is getting plenty of assistance these days, along with food sources such as certain fish and egg yolks. Many other food and drink products are vitamin D–fortified.

- **Supplements.** For boosting vitamin D levels, supplements do work, but be sure to read the labels. While tablets have posted up to 50,000 IUs, it is best to choose options within the daily-recommended dosage range, 600 to 800 IUs, says Yale endocrinologist Dr. Thomas Carpenter.[51]

- **Supplemental information.** Dr. Mike Ren of the Baylor College of Medicine suggests that your body may absorb only about one-sixth of the amount of vitamin D in a supplement. But don't worry about taking too much, he adds, because, "If there is a little excess, your body will excrete it out."[52]

ADDENDUM
A CHILLING TALE

Despite the ingenuity of Antarctic explorers Paul Siple and Charles Passel, and the scientists who have sought to perfect the wind-chill index with the help of human subjects, no measure can capture fully how the cold feels on an individual human body.

But in his 1898 short story, "The Cold Snap," the writer Edward Bellamy chillingly and descriptively addresses the "how does it feel" question in a way that transcends quantification.

For anyone who has experienced a period of extreme cold, this account of the terrors of a frigid spell endured by a family in New England should resonate.

Fiction can be truer than truth, an observation that generations of novelists and short story writers hold to be self-evident.

Granted, New England is not Antarctica, but as we've said, thermal comfort is relative, and on a night when the frigid air penetrates to our very bones, what would it matter to you if it's colder somewhere else?

Here are excerpts from Bellamy's story, courtesy of Project Gutenberg.[1]

+ + +

I have to take my vacations as the fluctuations of a rather exacting business permit, and so it happened that I was, with my wife, passing a fortnight in the coldest part of winter at the family homestead in New England. The ten previous days had been very cold, and the cold had "got into the house," which means that it had so penetrated and chilled the very walls and timbers that a cold day now took hold of us as it had not earlier in the season. Finally, there came a day that was colder than any before it. . . .

Toward dusk I took a short run to the post-office. I was well wrapped up, but that did not prevent me from having very singular sensations before I got home. The air, as I stepped out from cover, did not seem like air at all, but like some almost solid medium, whose impact was like a blow. It went right through my overcoat at the first assault, and nosed about hungrily for my little spark of vital heat. A strong wind with the flavor of glaciers was blowing straight from the pole. How inexpressibly bleak was the aspect of the leaden clouds that were banked up around the horizon! I shivered as I looked at the sullen masses. The houses seemed little citadels against the sky. I had not taken fifty steps before my face stiffened into a sort of mask, so that it hurt me to move the facial muscles. . . . Did the Creator intend man to inhabit high latitudes?

At nightfall father, Bill, and Jim, the two latter being my younger broth-ers, arrived from their offices, each in succession declaring, with many "whews" and "ughs," that it was by all odds the coldest night yet. . . .

After tea Ella [a sister] got the evening paper out of somebody's overcoat. . . . She read aloud: . . . Another cold wave on the way East. It will probably reach the New England states this evening. The thermometers along its course range from 40 degrees below zero at Fort Laramie, to 38 below in Omaha, 31 below in Chicago, and 30 degrees in Cleveland. Numerous cases of death by freezing are reported.

A gentleman friend called to take Ella out to a concert or something of the sort. Her mother was for having her give it up on account of the cold. But it so happens that young people, who, having life before them, can much better afford than their elders to forego particular pleasures, are much less resigned to doing so. The matter was compromised by piling so many wraps upon her that she protested it was like being put to bed. But, before they had been gone fifteen minutes, they were back again, half frozen. It had proved so shockingly cold they had not dared to keep on. . . . The streets were entirely deserted; not even a policeman was visible, and the chilled gas in the street lamps gave but a dull light.

Ella proposed to give us our regular evening treat of music, but found the corner of the room where the melodeon stood too cold. . . .

My wife is from the Southern states; and the huge cold of the North had been a new and rather terrifying experience to her. She had been growing nervous all the evening, as the signs and portents of the weather accumulated. She was really half frightened.

"Aren't you afraid it will get so cold it will never be able to get warm again—and then what would become of us?" she asked.

Of course, we laughed at her, but I think her fears infected me with a slight, vague anxiety, as the evidences of extraordinary and still increasing cold went on multiplying.

At length, one by one, the members of the family, with an anticipatory shiver . . . went to their rooms, and were doubtless in bed in the shortest possible time, and I fear without saying their prayers. Finally, my wife suggested that we had better go before we got too cold to do so.

The bedroom was shockingly cold. . . . As I lay awake, I heard the sides of the house crack in the cold. "What," said I to myself with a shiver, "should I do if anything happened that required me to get up and dress again?" It seemed to me I should be capable of letting a man die in the next room for need of succor. . . . The last thing I remember before dropping off to sleep was solemnly prom- ising my wife never to trust ourselves North another winter. I then fell asleep and dreamed of the ineffable cold of the interstellar spaces, which the scientific people talk about.

The next thing I was sensible of was a feeling of the most utter discomfort I ever experienced. My whole body had become gradually chilled through. I could feel the flesh rising in goose pimples at every movement. What has happened? was my first thought. The bedclothes were all there, four inches of them, and to find myself shivering under such a pile seemed a reversal of the laws of nature. Shivering is an unpleasant operation at best and at briefest; but when one has shivered till the flesh is lame, and every quiver is a racking, aching pain, that is something quite different from any ordinary shivering. My wife was awake and in the same condition. What did I ever bring her to this terrible country for? She had been lying as still as possible for an hour or so, waiting till she should die or something; and feeling that if she stirred she should freeze, as water near the freezing point crystallizes when agitated. She said that when I had disturbed the clothes by any movement, she had felt like hating me. . . .

I have had my share of unpleasant duties to face in my life. I remember how I felt at Spottsylvania when I stepped up and out from behind a breast-work of fence rails, over which the bullets were whistling like hailstones, to charge the enemy. . . . But never did an act of my life call for so much of sheer will-power as stepping out of that comfortless bed into that freezing room. It is a general rule in getting up winter mornings that the air never proves so cold as was anticipated while lying warm in bed. But it did this time, probably because my system was deprived of all elasticity and power of reaction by being so thoroughly chilled. Hastily donning in the dark what was absolutely necessary, my poor wife and myself, with chattering teeth and prickly bodies, the most thoroughly demoralized couple in history, ran downstairs to the sitting-room.

Much to our surprise, we found the gas lighted and the other members of the family already gathered there, huddling over the register. I felt a sinking at the heart as I marked the strained, anxious look on each face, a look that asked what strange thing had come upon us. They had been there, they said, for some time. Ella, Jim, and Bill, who slept alone, had been the first to leave their beds. Then father and mother, and finally my wife and I, had followed. Soon after our arrival there was a fumbling at the door, and the two Irish girls, who help mother keep house, put in their blue, pinched faces. They scarcely waited an invitation to come up to the register.

The room was but dimly lighted, for the gas, affected by the fearful chill, was flowing slowly and threatened to go out. The gloom added to the depressing effect of our strange situation. Little was said. The actual occurrence of strange and unheard-of events excites very much less wonderment than the account of them written or rehearsed. Indeed, the feeling of surprise often seems wholly left out of the mental experience of those who undergo or behold the most prodigious catastrophes. The sensibility to the marvelous is the one of our faculties which is, perhaps, the soonest exhausted by a strain. Human nature takes naturally to miracles, after all. "What can it mean?" was the inquiry a dozen times on the lips of each one of us, but beyond that, I recall little that was said. . . .

After piling all the coal on the furnace, it would hold, the volume of heat rising from the register was such as to singe the clothes of those over it, while those waiting their turn were shivering a few feet off. The men of course yielded the nearest places to the women, and, as we walked briskly up and down in the room, the frost gathered on our mustaches. *The morning, we said, would bring relief, but none of us fully believed it, for the strange experience we were enduring appeared to imply a suspension of the ordinary course of nature.* . . .

Swiftly it grew colder. The iron casing of the register was cold in spite of the volume of heat pouring through it. Every point or surface of metal in the room was covered with a thick coating of frost. The frost even settled upon a few filaments of cobweb in the corners of the room which had escaped the housemaid's broom, and which now shone like hidden sins in the day of judgment. The door-knob, mop-boards, and wooden casings of the room glistened. We were so chilled that woolen was as cold to the touch as wood or iron. Every few moments the beams of the house snapped like the timbers of a straining ship, and at intervals the frozen ground cracked with a noise like cannon. . . .

The cold is a sad enemy to beauty. My poor wife and Ella, with their pinched faces, strained, aching expression, red, rheumy eyes and noses, and blue or pallid cheeks were sad parodies on their comely selves. Other forces of nature have in them something the spirit of man can sympathize with, as the wind, the waves, the sun; but there is something terribly inhuman about the cold. I can imagine it as a congenial principle brooding over the face of chaos in the eons before light was.

Hours had passed, it might have been years, when father said, "Let us pray." He knelt down, and we all mechanically followed his example, as from childhood up we had done at morning and evening. Ever before, the act had seemed merely a fit and graceful ceremony, from which no one had expected anything in particular to follow, or had experienced aught save the placid reaction that commonly results from a devotional act. But now the meaning so long latent became eloquent. The morning and evening ceremony became the sole resource in an imminent and fearful emergency. There was a familiar strangeness about the act under these circumstances which touched us all. With me, as with most, something of the feeling implied in the adage, "Familiarity breeds contempt," had impaired my faith in the practical efficacy of prayer. How could extraordinary results be expected from so common an instrumentality, and especially from so ordinary and every-day a thing as family prayer? Our faith in the present instance was also not a little lessened by the peculiar nature of the visitation. In any ordinary emergency God might help us, but we had a sort of dim apprehension that even He could not do anything in such weather. . . . There are some inflictions which, although terrible, are capable of stirring in haughty human hearts a rebellious indignation. But to cold succumb soul and mind. It has always seemed to me that cold would have broken down Milton's Satan. . . .

Owing to the sustaining power there is in habit, the participation in family devotions proved strengthening to us all. In emergencies, we get back from our habits the mental and moral vigor that first went to their formation, and has since remained on interest.

It is not the weakest who succumb first to cold, as was strikingly proved in our experience. The prostration of the faculties may be long postponed by the power of the will. All assaults on human nature, whether of cold, exhaustion, terror, or any other kind, respect the dignity of the mind, and await its capitulation before finally storming the stronghold of life. . . .

The next thing I knew, Bill was urging us to eat some beefsteak and bread. The former, I afterward learned, he had got out of the pantry and cooked over the furnace fire. It was about five o'clock, and we had eaten nothing for nearly twelve hours. The general exhaustion of our powers had prevented a natural appetite from making itself felt, but mother had suggested that we should try food, and it saved us. It was still fearfully cold, but the danger was gone as soon as we felt the reviving effect of the food. An ounce of food is worth a pound of blankets. Trying to warm the body from the outside is working at a tremendous disadvantage. It was a strange picnic as, perched on chairs and tables in the dimly lighted room, we munched our morsels, or warmed the frozen bread over the register. After this, some of us got a little sleep.

I shall never forget my sensations when, at last, I looked out at the eastern window and saw the rising sun. The effect was indeed peculiarly splendid, for the air was full of particles of ice, and the sun had the effect of shining through a mist of diamond dust. Bill had dosed us with whiskey, and perhaps it had got into our heads, for I shouted, and my wife cried. It was, at the end of the weary night, like the first sight of our country's flag when returning from a foreign world.

ACKNOWLEDGMENTS

In undertaking such an immense and fascinating topic as biometeorology, I was beyond fortunate to have access to so many outstanding medical professionals who were ever so willing to share their wisdom.

I am indebted to Dr. Norman E. Rosenthal, who turned one of his darker moments into a monumental insight into how our moods and well-being are inextricably tied to the natural environment. His work on seasonal affective disorder profoundly changed how people relate to fall and winter. Over the years, he always has found time to answer my calls and notes and has been unsparingly generous with his insights.

I thank Dr. Estelle Levetin, emeritus professor at the University of Tulsa and master of aeroallergens, who has taught me so much about the chaotic world of pollen that I told her I properly should be sending her a tuition check. I also thank allergist Dr. Donald Dvorin, one of the relatively few remaining National Allergy Bureau genuine pollen counters, who never hesitated to share the knowledge he has accumulated in more than 30 years of practice and pollen analysis.

For what I have learned about the deadly effects of heat and its underrated role as the world's number-one weather killer, I am grateful to Dr. Laurence Kalkstein, Dr. Haresh Mirchandani, and Dr. Larry Robinson, who helped educate the nation and the world about heat hazards. Also, I thank Eric Klinenberg, the sociologist whose seminal *Heat Wave: A Social Autopsy of Disaster in Chicago* is essential reading for journalists or anyone else who wants to understand heat waves and the social and environmental forces likely to make them deadlier.

Harvard University's Dr. Paul G. Mathew was a treasury of information about headaches—particularly migraines—and their possible weather connections.

I am indebted to the works of biometeorology pioneers who revived interest in biometeorology in the mid-twentieth century, particularly Helmut E. Landsberg and Solco Tromp, who founded the International Society of Biometeorology in 1956.

I thank the Historical Medical Library at the Philadelphia College of Physicians for access to the papers of Dr. Joseph Hollander, who conducted one of the grand experiments in medical history to determine how weather might affect arthritis.

The American Academy of Allergy, Asthma & Immunology, the National Headache Foundation, and the National Association of School Nurses were among other valuable resources.

This book would have remained a concept without the patient encouragement from Prometheus editor Jake Bonar; the gentle prodding from my agent, Anne Devlin; and the support I have received from long-suffering *Inquirer* editors Molly Eichel, Emily Babay, Diane Mastrull, and the late Kathleen Hacker.

For her assistance with the notes, I owe Allison Outlund my sanity, and I thank Veronica Jurgena for her careful copyediting, a tremendously underrated skill.

And most of all I am grateful to my ever-supportive wife, Laura, who has endured my atmo-chondria through the years while insisting that other things beyond the atmosphere govern our health.

I have taken her perspective under advisement.

NOTES

PREFACE

1. Committee on Climate, Ecosystems, Infectious Disease, and Human Health and National Research Council Staff, "Preface," in *Under the Weather: Climate, Ecosystems, and Infectious Disease* (Washington, DC: National Academies Press, May 2001), p. xii, https://nap.nationalacademies.org/read/10025/chapter/1#xi.

2. Glenville Jones, "100 Years of Vitamin D: Historical Aspects of Vitamin D," *Endocrine Connections* 11, no. 4 (April 22, 2022): e210594, https://doi.org/10.1530/EC-21-0594.

INTRODUCTION

1. National Academies of Sciences, Engineering, and Medicine and National Research Council Staff, *Under the Weather: Climate, Ecosystems, and Infectious Disease* (Washington, DC: National Academies Press, 2001), p. 1.

2. James H. Cassedy, *Medicine and American Growth, 1800–1860* (Madison: University of Wisconsin Press, 1986), pp. 44–45, 48.

3. Alan E. Stewart, "Edwin Grant Dexter: An Early Researcher in Human Behavioral Biometeorology," *International Journal of Biometeorology* 59, no. 6 (June 2015): 745–58, https://doi.org/10.1007/s00484-014-0888-3.

4. Edwin Grant Dexter, *Weather Influences: An Empirical Study of the Mental and Physiological Effects of Definite Meteorological Conditions* (London: MacMillan Co., 1904), Preface, Google Books, https://www.google.com/books/edition/Weather_Influences/se_UAAAAMAAJ?hl=en&gbpv=1.

5. René Habert, "Claude Bernard, the Founder of Modern Medicine," *Cells* 11, no. 10 (2022): 1702, https://www.ncbi.nlm.nih.gov/pmc/articles/PMC9139283/.

6. Frederick Sargent II, "The Nature and Nurture of Biometeorology," *Bulletin of the American Meteorological Society* 13, no. 3 (June 1963): 20–23, https://doi.org/10.2307/1293081.

7. Hippocrates, *On Airs, Waters, and Places*, trans. Francis Adams (London: Wyman & Sons, 1881), p. 3.

8. Eleni Tsiompanou and Spyros G. Marketos, "Hippocrates: Timeless Still," *Journal of the Royal Society of Medicine* 106, no. 7 (2013): 288–92, https://www.ncbi.nlm.nih.gov/pmc/articles/PMC3704070/.

9. Fielding H. Garrison, *Introduction to the History of Medicine*, 4th ed. (Philadelphia: W.B. Saunders Co., 1929), pp. 92–93.

10. Ann Ellis Hanson, "Hippocrates: The 'Greek Miracle' in Medicine," Medicina Antiqua, accessed May 31, 2024, https://www.ucl.ac.uk/~ucgajpd/medicina%20antiqua/sa_hippint.html#:~:text=The%20Hippocratic%20Corpus%20consists%20of,of%20the%20Athenian%20democracy%2C%20when.

11. Tsiompanou and Marketos, "Hippocrates: Timeless Still."

12. "Olympian Healers," Ancient Greek Medicine, National Library of Medicine, accessed May 31, 2024, https://www.nlm.nih.gov/hmd/topics/greek-medicine/index.html#case1.

13. Tsiompanou and Marketos, "Hippocrates: Timeless Still."

14. Tsiompanou and Marketos, "Hippocrates: Timeless Still."

15. Garrison, *History of Medicine*, p. 92.

16. Hippocrates, *On Airs, Water and Places*, p. 3.

17. National Academies of Sciences, Engineering, and Medicine and National Research Council Staff, *Under the Weather*, p. 8.

18. "Facts about the Common Cold," Lung Health & Diseases, American Lung Association, updated January 22, 2024, https://www.lung.org/lung-health-diseases/lung-disease-lookup/facts-about-the-common-cold#:~:text=Rhinovirus%20is%20the%20most%20common,and%20those%20in%20poor%20health.

19. National Academies of Sciences, Engineering, and Medicine and National Research Council Staff, *Under the Weather*, p. 14.

20. National Academies of Sciences, Engineering, and Medicine and National Research Council Staff, *Under the Weather*, p. 14.

21. Jay Lawrimore, "Thomas Jefferson and the Telegraph: Highlights of the U.S. Weather Observer Program," Climate.gov, May 17, 2018, https://www.climate.gov/news-features/blogs/beyond-data/thomas-jefferson-and-telegraph-highlights-us-weather-observer.

22. Cassedy, *Medicine and American Growth*, pp. 44–45, 48.

23. National Academies of Sciences, Engineering, and Medicine and National Research Council Staff, *Under the Weather*, pp. 14–15.

24. "Introduction to the National Weather Service," National Weather Service, National Oceanic and Atmospheric Administration, accessed May 31, 2024, https://w2.weather.gov/jetstream/nws_intro.

25. National Academies of Sciences, Engineering, and Medicine and National Research Council Staff, *Under the Weather*, pp. 15–16.

26. Michael A. Osborne and Richard S. Fogarty, "Medical Climatology in France: The Persistence of Neo-Hippocratic Ideas in the First Half of the Twentieth Century," *Bulletin of the History of Medicine* 86, no. 4 (Winter 2012): 543–63, https://www.jstor.org/stable/26305890.

27. G. Edgar Folk, *The International Society of Biometeorology: A Fifty-Year History*, January 1997, https://uwm.edu/biometeorology/wp-content/uploads/sites/439/2017/06/ISB_50YearHistoryofBiometeorology.pdf.

28. Folk, *International Society of Biometeorology*, p. 8.

29. Ogone Motlogeloa and Jennifer M. Fitchett, "Climate and Human Health: A Review of Publication Trends in the International Journal of Biometeorology," *International Journal of Biometeorology* 67 (May 2023): 933–55, https://doi.org/10.1007/s00484-023-02466-8.

30. Charles W. Schmidt, "Pollen Overload: Seasonal Allergies in a Changing Climate," *Environmental Health Perspectives* 124, no. 4 (April 2016): A70–A75, https://doi.org/10.1289/ehp.124-A70.

SPRING

1. Christie Nicholson, "Fact or Fiction? 'Spring Fever' Is a Real Phenomenon," *Scientific American*, March 22, 2007, https://www.scientificamerican.com/article/fact-or-fiction-spring-fever-is-a-real-phenomenon/.

2. Jeanie Lerche Davis, "You Give Me (Spring) Fever," WebMD, medically reviewed by Sabrina Felson, April 27, 2022, https://www.webmd.com/women/features/you-give-me-spring-fever.

3. Timothy M. Smith, "Sleep Doctors' Orders: Use Standard Time 365 Days a Year," American Medical Association, March 5, 2024, https://www.ama-assn.org/delivering-care/public-health/sleep-doctors-orders-use-standard-time-365-days-year.

CHAPTER 1

1. Donald Dvorin, interview by the author, March 21, 2023.

2. "National Allergy Bureau," American Academy of Allergy, Asthma & Immunology, accessed June 3, 2024, https://pollen.aaaai.org/#/.

3. Mayo Clinic Staff, "Allergies and Asthma: They Often Occur Together," Mayo Clinic, May 14, 2021, https://www.mayoclinic.org/diseases-conditions/asthma/in-depth/allergies-and-asthma/art-20047458#:~:text=The%20same%20substances%20that%20trigger,asthma%20or%20allergy%2Dinduced%20asthma.

4. "Rising CO2, Climate Change and Allergies," News & Events, Columbia Mailman School of Public Health, Columbia University Irving Medical Center, accessed June 3, 2024, https://www.publichealth.columbia.edu/research/centers/niehs-center-environmental-health-justice-northern-manhattan/news-events/newsbriefs/fall-2020/rising-co2-climate-change-allergies#:~:text=Increasing%20carbon%20dioxide%20in%20the,where%20these%20allergens%20are%20found.

5. Diane E. Pataki, Marina Alberti, Mary Cadenasso, Alexander Felson, Mark J. McDonnell, Stephanie Pincetl, Richard V. Pouyat, Heikki Setälä, and Thomas H. Whitlow, "The Benefits and Limits of Urban Tree Planting for Environmental and Human Health," *Frontiers in Ecology and Evolution* 9 (2021): 603757, https://doi.org/10.3389/fevo.2021.603757.

6. Daniel S. W. Katz, "Effect of Intra-Urban Temperature Variation on Tree Flowering Phenology, Airborne Pollen, and Measurement Error in Epidemiological Studies of Allergenic Pollen," *Science of the Total Environment* 653 (2019): 1213–22, https://doi.org/10.1016/j.scitotenv.2018.11.020.

7. Charles W. Schmidt, "Pollen Overload: Seasonal Allergies in a Changing Climate," *Environmental Health Perspectives* 124, no. 4 (2016): A70–A75, https://doi.org/10.1289/ehp.124-A70.

8. "Climate and Health," Centers for Disease Control and Prevention, March 2, 2024, https://www.cdc.gov/climate-health/php/effects/allergens-and-pollen.html#:~:text=Medical%20costs%E2%80%8E,be%20more%20sensitive%20to%20pollen.

9. J. M. S. Pearce, "John Bostock and Hay Fever," Hektoen International, August 25, 2022, https://hekint.org/2022/08/25/john-bostock-and-hay-fever/.

10. John Bostock, "Case of a Periodical Affection of the Eyes and Chest," *Medico-Chirurgical Transactions of London* 10, part 1 (1819):161–65, https://www.ncbi.nlm.nih.gov/pmc/articles/PMC2116437/pdf/medcht00109-0169.pdf.

11. Pearce, "John Bostock."

12. William Heberden, *Commentaries on the History and Cure of Diseases* (Boston: Wells and Lilly, 1818), p. 109, Internet Archive, https://archive.org/details/commentariesonhi00heberich/page/n9/mode/2up?view=theater.

13. Heberden, *Commentaries*, p. v.

14. Manoj Ramachandran and Jeffrey K. Aronson, "John Bostock's First Description of Hayfever," *Journal of the Royal Society of Medicine* 104, no. 6 (2011): 237–40, https://doi.org/10.1258/jrsm.2010.10k056.

15. Donald W. Cockroft, "Allergen-Induced Asthma," *Canadian Respiratory Journal* 21, no. 5 (2014): 279–82, https://doi.org/10.1155/2014/719272.

16. Cockroft, "Allergen-Induced Asthma."

17. Harold S. Nelson, "The Evolution of Allergy Immunotherapy," *Annals of Allergy, Asthma & Immunology* 126, no. 4 (April 2021): 357–66, https://doi.org/10.1016/j.anai.2020.11.011.

18. Morrill Wyman, "Autumnal Catarrh," *Boston Medical and Surgical Journal* 93, no. 8 (August 19, 1875): 209–12, https://doi.org/10.1056/NEJM187508190930801.

19. Wyman, "Autumnal Catarrh."

20. Julian Crane, "Charles Harrison Blackley: The Man Who Put the Hay in Hay Fever," Hektoen International, Fall 2015, https://hekint.org/2017/01/28/charles-harrison-blackley-the-man-who-put-the-hay-in-hay-fever/.

21. Crane, "Charles Harrison Blackley."

22. Crane, "Charles Harrison Blackley."

23. Thomas A. E. Platts-Mills, "The Allergy Epidemics: 1870–2010," *Journal of Allergy and Clinical Immunology* 136, no. 1 (2015): 3–13, https://doi.org/10.1016/j.jaci.2015.03.048.

24. Matthew Walzer and Bernard B. Siegel, "The Effectiveness of the Ragweed Eradication Campaigns in New York City: A 9-Year Study (1946–1954)," *Journal of Allergy* 27, no. 2 (March 1956): 113–26, https://doi.org/10.1016/0021-8707(56)90002-8.

25. M. C. Matheson, E. H. Walters, J. A. Simpson, C. L. Wharton, A.-L. Ponsonby, D. P. Johns, M. A. Jenkins, G. G. Giles, J. L. Hopper, M. J. Abramson, and S. C. Dharmage, "Relevance of the Hygiene Hypothesis to Early vs. Late Onset Allergic Rhinitis," *Clinical & Experimental Allergy* 39, no. 3 (March 2009): 370–78, https://doi.org/10.1111/j.1365-2222.2008.03175.x.

26. Carmella Wint, "Everything You Need to Know About Sneezing," updated July 20, 2019, https://www.healthline.com/health/sneezing#medical-treatment.

27. Lydia Bourouiba, Eline Dehandschoewercker, and John W. M. Bush, "Violent Expiratory Events: On Coughing and Sneezing," *Journal of Fluid Mechanics* 745 (March 2014): 537–63, https://doi.org/10.1017/jfm.2014.88.

28. Jenny Adams, Jack Schmid, Robert D. Parker, J. Richard Coast, Dunlei Cheng, Aaron D. Killian, Stephanie McCray, Danielle Strauss, Sandra McLeroy DeJong, and·Rafic Berbarie, "Comparison of Force Exerted on the Sternum During a Sneeze Versus During Low-, Moderate-, and High-Intensity Press Resistance Exercise with and without the Valsalva Maneuver in Healthy Volunteers," *American Journal of Cardiology* 113, no. 6 (March 15, 2014): 1045–48, https://doi.org/10.1016/j.amjcard.2013.11.064.

29. Luis Villazon, "Why Do We Make the 'Atchoo' Sound When We Sneeze?," BBC Science Focus, accessed June 3, 2024, https://www.sciencefocus.com/the-human-body/why-do-we-make-the-atchoo-sound-when-we-sneeze.

30. D. P. Strachan, "Hay Fever, Hygiene, and Household Size," *British Medical Journal* 299 (November 1989): 1259–60, https://doi.org/10.1136/bmj.299.36710.1259.

31. Platts-Mills, "The Allergy Epidemic."

32. Asthma and Allergy Foundation of America, "Asthma and Allergy Foundation of America Announces 2023 Allergy Capitals," March 15, 2023, https://aafa.org/asthma-and-allergy-foundation-of-america-announces-2023-allergy-capitals/.

33. William R. L. Anderegg, John T. Abatzoglou, Leander D. L. Anderegg, and Lewis Ziska, "Anthropogenic Climate Change Is Worsening North American Pollen Seasons," *Proceedings of the National Academy of Sciences* 118, no. 7 (February 16, 2021): e2013284118, https://doi.org/10.1073/pnas.2013284118.

34. Anthony R. Wood, "Tree Pollen Hits 'Extreme' Level as the Season Starts in the Philly Region," *Philadelphia Inquirer*, March 22, 2023, https://www.inquirer.com/news/pollen-counts-allergies-philadelphia-weather-climate-20230322.html.

35. Beverley Adams-Groom, Katherine Selby, Sally Derrett, Carl A. Frisk, Catherine Helen Pashley, Jack Satchwell, Dale King,Gaynor McKenzie, and Roy Neilson, "Pollen Season Trends as Markers of Climate Change Impact: *Betula, Quercus* and *Poaceae*," *Science of the Total Environment* 831 (July 2022): 154882, https://doi.org/10.1016/j.scitotenv.2022.154882.

36. A. B. Singh and Pawan Kumar, "Climate Change and Allergic Diseases: An Overview," *Frontiers in Allergy* 3 (October 13, 2022): 964987, https://doi.org/10.3389/falgy.2022.964987.

37. "Featured Member: Fiona Lo," University of Washington Department of Global Health, Center for Health and the Global Environment, accessed June 3, 2024, https://deohs.washington.edu/change/featured-member-fiona-lo.

38. Estelle Levetin, email message to author, August 12, 2020.

39. Wood, "Tree Pollen Hits 'Extreme' Level."

40. Estelle Levitin, email message to author, March 30, 2011.

41. Mike Doll (senior meteorologist, AccuWeather) in discussion with the author, April 5, 2022.

42. Matthew Rank, Divya Shah, Linda Ford, Sergei Ochkur, Susan Kosisky, Stanley Fineman, and Frank Virant, "Accuracy of Spring Pollen Forecasts in Five United States Cities Using National Allergy Bureau Reporting as a Gold Standard," *Journal of Allergy and Clinical Immunology* 153, no. 2 (February 2024): AB109, https://doi.org/10.1016/j .jaci.2023.11357.

43. Anthony R. Wood, "Philly Region's Dry Spell Intensifies, and the Tree Pollen Is Loving It More than Allergy Sufferers," *Philadelphia Inquirer*, April 21, 2023, https: //www.inquirer.com/weather/philadelphia-weather-pollen-counts-forecast-20230420 .html?utm_source=headtopics&utm_medium=news&utm_campaign=2023-04-21.

44. Dvorin, interview by the author, March 21, 2023.

45. Anthony R. Wood, "As Pollen Torments Millions, It Might Be Getting Worse, and It's Poorly Measured in America," *Philadelphia Inquirer*, May 8, 2021, https:// www.inquirer.com/news/pollen-allergies-coronavirus-philadelphia-trees-grasses-climate -change-asthma-20210508.html.

46. Estelle Levitin, "Use of the Burkard Spore Trap," American Academy of Allergy, Asthma and Immunology, accessed June 3, 2024, https://education.aaaai.org/sites/default /files/Burkard%20Directions%20Handout-1.pdf.

47. Wood, "As Pollen Torments Millions."

48. Fiona Lo, email message to author, April 26, 2021.

49. Fiona Lo, email message to author, April 27, 2021.

50. "What Does It Mean to Be a Certified Pollen Counter?," The Allergy Group, April 24, 2017, https://theallergygroup.com/what-does-it-mean-to-be-a-certified-pollen -counter/.

51. "National Allergy Bureau."

52. Yolanda Clewlow, email message to author, May 5, 2021.

53. Jeroen Buters, email message to author, May 4, 2011.

54. J. Buters, C. Antunes, A. Galveias, K. C. Bergmann, M. Thibaudon, C. Galán, C. Schmidt-Weber, and J. Oteros, "Pollen and Spore Monitoring in the World," *Clinical and Translational Allergy* 8 (2018): 9, https://doi.org/10.1186/s13601-018-0197-8.

55. Buters et al., "Pollen and Spore Monitoring in the World."

56. Fiona Lo, email message to author, April 30, 2021.

57. "Message to Our Loyal Followers," University of Tulsa Mountain Cedar Pollen Forecasts, accessed June 3, 2024, http://pollen.utulsa.edu/current.html.

58. Jakob Schaefer, Manuel Milling, Björn W. Schuller, Bernhard Bauer, Jens O. Brunner, Claudia Traidl-Hoffmann, and Athanasios Damialis, "Towards Automatic Airborne Pollen Monitoring: From Commercial Devices to Operational by Mitigating Class-Imbalance in a Deep Learning Approach," *Science of the Total Environment* 796 (November 20, 2021): 148932, https://doi.org/10.1016/j.scitotenv.2021.148932.

59. Fiona Lo, email message to author, April 30, 2021.

CHAPTER 2

1. Joseph L. Hollander and Sarantos J. Yeostros, "The Effect of Simultaneous Variations of Humidity and Barometric Pressure on Arthritis," *Bulletin of the American Meteorological Society* 44, no. 1 (August 1963): 489–94.

2. Joseph L. Hollander, ed., *Arthritis and Allied Conditions* (Philadelphia: Lea & Febiger, 1966), p. 21.

3. Linda Rath, "What Is Arthritis?," Arthritis Foundation, updated June 9, 2022, https://www.arthritis.org/health-wellness/about-arthritis/understanding-arthritis/what-is-arthritis.

4. "The Joseph Lee Hollander Professorship of Pediatric Rheumatology," Perelman School of Medicine, University of Pennsylvania, accessed May 28, 2024, https://www.med.upenn.edu/endowedprofessorships/joseph-lee-hollander-professorship-of-pediatric-rheumatology.html.

5. Robert Eisenberg, H. Ralph Schumacher, Terri H. Finkel, and George E. Ehrlich, "In Memoriam, Joseph Lee Hollander," *Arthritis & Rheumatism* 43, no. 27 (July 2000): 1430, https://doi.org/10.1002/1529-0131(200007)43:7.

6. Charlotte H. Anderson, "11 Ways to Describe Arthritis So Other People Get What It Really Feels Like," CreakyJoints.org, last modified January 30, 2019, https://creakyjoints.org/support/what-arthritis-really-feels-like/.

7. Anthony R. Wood, "A Mix of Rain and Pain," *Philadelphia Inquirer,* April 22, 2022, B1, B3.

8. N. Iikuni, A. Nakajima, E. Inoue, E. Tanaka, H. Okamoto, M. Hara, T. Tomatsu, N. Kamatani, and H. Yamanaka, "What's in Season for Rheumatoid Arthritis Patients? Seasonal Fluctuations in Disease Activity," *Rheumatology* (May 2007): 846–48, https://pubmed.ncbi.nlm.nih.gov/17264092/.

9. Hollander and Yeostros, "Effect of Simultaneous Variations," p. 489.

10. Hollander and Yeostros, "Effect of Simultaneous Variations," p. 494.

11. "Dew Point vs Humidity," National Weather Service, accessed May 28, 2024, https://www.weather.gov/arx/why_dewpoint_vs_humidity.

12. "Air Pressure," National Oceanic and Atmospheric Administration, updated December 18, 2023, https://www.noaa.gov/jetstream/atmosphere/air-pressure.

13. "The Top of Mt. Everest Sits," School of Chemistry, University of Bristol, accessed May 28, 2024, https://www.chm.bris.ac.uk/webprojects2004/white/top_of_mount_everest_sits_in_the.htm.

14. James Joyce, *Counterparts* (London: Grant Richards Ltd., 1914), p. 109.

15. Charles Jackson, *The Lost Weekend* (New York: Farrar, Straus & Giroux, 1965), https://gutenberg.ca/ebooks/jacksoncr-lostweekend/jacksoncr-lostweekend-00-h.html.

16. John B. West, "Torricelli and the Ocean of Air: The First Measurement of Barometric Pressure." *Physiology* (Bethesda, MD) 28, no. 2 (March 2013): 66–73, https://doi.org/10.1152/physiol.00053.2012.

17. Deborah Lynn Blumberg, "Does Weather Affect Joint Pain?," Pain Management, WebMD, reviewed by David Zelman November 27, 2022, https://www.webmd.com/pain-management/weather-and-joint-pain.

18. Robert H. Shmerling, "Can the Weather Really Worsen Arthritis Pain?," *Harvard Health* (blog), Harvard Health Publishing, June 22, 2020, https://www.health.harvard.edu/blog/can-the-weather-really-worsen-arthritis-pain-201511208661.

19. "Hall of Fame Members," American Society of Heating, Refrigerating, and Air-Conditioning Engineers, accessed September 30, 2024, https://www.ashrae.org/membership/honors-and-awards/hall-of-fame-members.

20. John Everetts Jr., "Design of Climate-Control Chamber," *Transactions of the New York Academy of Sciences* 24, no. 2, series II (December 1961), https://doi.org/10.1111/j.2164-0947.1961.tb00761.x.

21. Hollander and Yeostros, "Effect of Simultaneous Variations," p. 491.

22. Hollander and Yeostros, "Effect of Simultaneous Variations," p. 493.

23. Hollander, *Arthritis and Allied Conditions*, 489–94.

24. "Bomb," Glossary, National Weather Service, National Oceanic and Atmospheric Administration, accessed May 28, 2024, https://w1.weather.gov/glossary/index.php?word=bomb.

25. John Gyakum, interview with the author, May 23, 2024.

26. David M. Schultz, Anna L. Beukenhorst, Belay Birlie Yimer, Louise Cook, Huai Leng Pisaniello, Thomas House, Carolyn Gamble, Jamie C. Sergeant, John McBeth, and William G. Dixon, "Weather Patterns Associated with Pain in Chronic-Pain Sufferers," *Bulletin of the American Meteorological Society* (May 1, 2020): e555–e565, https://journals.ametsoc.org/view/journals/bams/101/5/bams-d-19-0265.1.xml.

27. Shmerling, "Can the Weather Really Worsen Arthritis Pain?"

28. Shmerling, "Can the Weather Really Worsen Arthritis Pain?" (emphasis mine).

29. Robert H. Shmerling, email message to author, May 15, 2024.

30. "Mapping the Weather Patterns Affecting People with Chronic Pain," University of Manchester, accessed May 25, 2024, https://www.manchester.ac.uk/discover/news/mapping-the-weather-patterns-affecting-people-with-chronic-pain/.

31. William C. Shiel Jr., "Whether Weather Affects Arthritis," MedicineNet, accessed May 28, 2024, https://www.medicinenet.com/arthritis_-_whether_weather_affects_arthritis/views.htm.

32. Shiel, "Whether Weather Affects Arthritis."

33. Shiel, "Whether Weather Affects Arthritis."

34. Anthony R. Wood, "Under the Weather," *Philadelphia Inquirer*, June 23, 2002, A-1, A-14.

35. Roy Gibbons, "Gramps Aches Good Barometer After All," *Chicago Tribune*, June 22, 1962, p. 1.

36. Karl Ubell, "Science Confirms Old Weather Adage," *Rochester Democrat and Chronicle*, July 1, 1962, p. 17.

37. "4 Tips for Managing Chronic Pain," Arthritis Foundation, accessed May 13, 2024, https://www.arthritis.org/health-wellness/healthy-living/managing-pain/pain-relief-solutions/4-tips-for-managing-chronic-pain.

CHAPTER 3

1. Dr. Paul Mathew, interview by the author, November 13, 2023.

2. "I Am an Old Man and Have Known a Great Many Troubles, But Most of Them Never Happened," Quote Investigator, October 4, 2013, https://quoteinvestigator.com /2013/10/04/never-happened/.

3. Paul Mathew, email follow-up comments to author, May 11, 2024.

4. Mathew, email follow-up comments.

5. "Headaches," Cleveland Clinic, last modified August 29, 2022, https://my .clevelandclinic.org/health/diseases/9639-headaches.

6. Joseph V. Campellone, "Headache," Medline Plus, last modified December 31, 2023, https://medlineplus.gov/ency/article/003024.htm.

7. Arne May, "Hints on Diagnosing and Treating Headache," *Deutsches Arzteblatt International* 115, no. 17 (April 27, 2018): 299–308, doi: 10.3238/arztebl.2018.0299.

8. "Part II: The Secondary Headaches," International Headache Society Classification, ICHD-3, International Headache Society, accessed May 17, 2024, https://ichd-3.org/.

9. Mathew, interview by the author, November 13, 2023.

10. "Weather Fronts," University Center for Atmospheric Research, accessed May 17, 2024, https://scied.ucar.edu/learning-zone/how-weather-works/weather-fronts.

11. Mathew, interview by the author.

12. Rebecca Buffum Taylor, "Can Changes in Weather Trigger Migraine and Other Headaches?" WebMD, September 22, 2023, https://www.webmd.com/migraines -headaches/headache-and-migraine-trigger-weather.

13. Jan Hoffmann, Tonio Schirra, Hendra Lo, Lars Neeb, Uwe Reuter, and Peter Martus, "The Influence of Weather on Migraine—Are Migraine Attacks Predictable?" *Annals of Clinical and Translational Neurology* 2, no. 1 (January 2, 2015): 22–28, https://doi.org /10.1002/acn3.139.

14. Marco A. Pescador Ruschel and Orlando De Jesus, "Epidemiology," in *Migraine Headache* (Treasure Island, FL: StatPearls Publishing, 2024), https://www.ncbi.nlm.nih .gov/books/NBK560787/#_article-22614_s4_.

15. Jatin Gupta and Sagar S. Gaurkar, "Migraine: An Underestimated Neurological Condition Affecting Billions," *Cureus* 14, no. 8 (August 24, 2022): e28347, https://doi .org/10.7759/cureus.28347.

16. Merriam-Webster Online Dictionary, s.v. "migraine (*n.*)," accessed June 12, 2024, https://www.merriam-webster.com/dictionary/migraine.

17. "What Type of Headache Do You Have?," American Migraine Foundation, January 19, 2023, https://americanmigrainefoundation.org/resource-library/what-type-of -headache-do-you-have/.

18. F. Clifford Rose, "The History of Migraine from Mesopotamian to Medieval Times," *Cephalalgia*, no. S15 (October 1995): 1–3, https://doi.org/10.1111/j.1468-2982 .1995.TB00040.X.

19. Mathew, email follow-up comments to author, May 11, 2024.

20. Dong-Gyun Han, "Evolutionary Game Model of Migraine Based on the Human Brain Hypersensitivity," *Frontiers in Neurology* 14 (March 28, 2023), https://doi.org/10 .3389/fneur.2023.1123978.

21. Mathew, email follow-up comments.

22. Mathew, email follow-up comments.

23. Mathew, email follow-up comments.

24. "Seasonal Migraine Triggers," American Migraine Foundation, May 13, 2021, https://americanmigrainefoundation.org/resource-library/seasonal-migraine/.

25. Hirohisa Okuma, Yumiko Okuma, and Yasuhisa Kitagawa, "Examination of Fluctuations in Atmospheric Pressure Related to Migraine," *Springerplus* 4 (December 18, 2015): 790, https://doi.org/10.1186/s40064-015-1592-4.

26. W. J. Becker, "Weather and Migraine: Can So Many Patients Be Wrong?" *Cephalalgia* 31, no. 4 (2011): 387–90, https://doi.org/10.1177/0333102410385583.

27. Masahito Katsuki, Muneto Tatsumoto, Kazuhito Kimoto, Takashige Iiyama, Masato Tajima, Tsuyoshi Munakata, Taihei Miyamoto, and Tomokazu Shimazu, "Investigating the Effects of Weather on Headache Occurrence Using a Smartphone Application and Artificial Intelligence," *Headache* 63, no. 5 (May 2023): 585–600, https://doi.org/10.1111/head.14482.

28. Jan Hoffmann, Hendra Lo, Lars Neeb, Peter Martus, and Uwe Reuter, "Weather Sensitivity in Migraineurs," *Journal of Neurology* 258, no. 4 (2011): 596–602, https://doi.org/10.1007/s00415-010-5798-7.

29. Stephanie Watson and Erika Klein, "Migraine and Weather Changes: What's the Link?" Healthline.com, last modified May 17, 2024, https://www.healthline.com/health/migraine/weather-connection#weather-migraine.

30. Albert C. Yang, J. L. Fuh, and N. E. Huang, "Patients with Migraine Are Right about Their Perception of Temperature as a Trigger," *Journal of Headache and Pain* 16 (2015): 49, https://doi.org/10.1186/s10194-015-0533-5.

31. Jörg Scheidt, Christina Koppe, Sven Rill, Dirk Reinel, Florian Wogenstein, and Johannes Drescher, "Influence of Temperature Changes on Migraine Occurrence in Germany," *International Journal of Biometeorology* 57 (2013): 649–54, https://doi.org/10.1007/s00484-012-0582-2.

32. Felix M. Key, Muslihudeen A. Abdul-Aziz, Roger Mundry, Benjamin M. Peter, Aarthi Sekar, Mauro D'Amato, Megan Y. Dennis, Joshua M. Schmidt, and Aida M. Andrés, "Human Local Adaptation of the TRPM8 Cold Receptor along a Latitudinal Cline," *Plos Genetics* (May 3, 2018), https://doi.org/10.1371/journal.pgen.1007298.

33. "Top 10 Migraine Triggers and How to Deal with Them," American Migraine Foundation, July 27, 2017, https://americanmigrainefoundation.org/resource-library/top-10-migraine-triggers/.

34. Mathew, email follow-up comments to author, May 11, 2024.

35. Harvard Health Publishing Staff, "Living with Chronic Headache: A Personal Migraine Story," Harvard Health Publishing, Harvard Medical School, March 5, 2011, https://www.health.harvard.edu/blog/living-with-chronic-headache-a-personal-migraine-story-201103051601.

36. "Barbara's Experience of Migraine with Aura," *The Migraine Trust*, accessed May 21, 2024, https://migrainetrust.org/understand-migraine/impact-of-migraine/barabaras-experience-of-migraine-with-aura/.

37. Carol Rääbus, "'I Spent Two Years Living in Bed': Your Stories of Living with Migraines," ABC News, last modified December 15, 2020, https://www.abc.net.au/everyday/your-stories-of-living-with-migraines/10743632.

38. "Both Sides of the Story (Part 1): Two Neurologists Specializing in Migraine Who Have Lived with the Disease for Decades," University of Utah School of Medicine, June 26, 2023, https://medicine.utah.edu/neurology/news/2023/06/both-sides-of-the-story -part-one.

39. Mathew, email follow-up comments.

40. Y.-Jung Lee, Y. T. Chen, S. M. Ou, S. Y. Li, A.C. Yang, C. H. Tang, and S. J. Wang, "Temperature Variation and the Incidence of Cluster Headache Periods: A Nationwide Population Study," *Cephalalgia* 34, no. 9 (August 2014): 656–663, https://pubmed.ncbi .nlm.nih.gov/24477598/.

41. Mathew, interview by the author.

42. "Light and Headache Disorders: Understanding Light Triggers and Photophobia," National Headache Foundation, accessed May 21, 2024, https: //headaches.org/light-headache-disorders-understanding-light-triggers -photophobia/#:~:text=During%20a%20series%2C%20cluster%20headache ,those%20without%20a%20headache%20disorder.

43. Ashley Hattle, "Why Are Cluster Headaches Affected by the Seasons?" Association of Migraine Disorders, March 27, 2020, https://www.migrainedisorders.org/cluster -headaches-seasonal-impact/#:~:text=Why%20cluster%20headaches%20are%20linked ,sleep%20cycles%20leading%20to%20attacks.

44. Mathew, email follow-up comments.

45. Becker, "Weather and Migraine."

46. H. Kesserwani, "Migraine Triggers: An Overview of the Pharmacology, Biochemistry, Atmospherics, and Their Effects on Neural Networks," *Cureus* 13, no. 4 (April 1, 2021): e14243, https://doi.org/10.7759/cureus.14243.

47. Becker, "Weather and Migraine."

48. "Chinook," Glossary, National Weather Service, accessed June 12, 2024, https:// forecast.weather.gov/glossary.php?word=CHINOOK.

49. Becker, "Weather and Migraine."

50. Stephen H. Schneider and Randi Londer, *The Coevolution of Climate and Life* (San Francisco: Sierra Club Books, 1984), p. viii.

51. Becker, "Weather and Migraine."

52. Mayo Clinic Staff, "Headaches: Treatment Depends on Your Diagnosis and Symptoms," Mayo Clinic, May 10, 2019, https://www.mayoclinic.org/diseases-conditions/ chronic-daily-headaches/in-depth/headaches/art-20047375.

ADDENDUM

1. Charles H. Blackley, *Hayfever: Its Causes, Treatment, and Effective Prevention*, 2nd edition (London: Bailliere, Tindall, & Sons, 1890), pp. 91–92.

SUMMER

1. Kelly Rohan, email message to author, August 1, 2023.

2. Tonya Ladipo, interview by the author, August 23, 2023.

3. Katarzyna Kliniec, Maciej Tota, Aleksandra Zalesińska, Magdalena Łyko, and Alina Jankowska-Konsu, "Skin Cancer Risk, Sun-Protection Knowledge and Behavior in Athletes—A Narrative Review," *Cancers (Basel)* 15, no. 13 (June 22, 2023): 3281, https://doi.org/10.3390/cancers15133281.

4. Dr. Seana Covello, interview by the author, February 28, 2023.

5. Brittany Behm, "Avoid Food Poisoning During Summer Picnics," *Public Health Matters Blog*, Centers for Disease Control and Prevention, July 10, 2017, https://blogs.cdc.gov/publichealthmatters/2017/07/avoid-food-poisoning-during-summer-picnics/.

6. Sean M. Moore, Rebecca J. Eisen, Andrew Monaghan, and Paul Mead, "Meteorological Influences on the Seasonality of Lyme Disease in the United States," *American Journal of Tropical Medicine and Hygiene* 90, no. 3 (March 5, 2014): 486–96, https://doi.org/10.4269/ajtmh.13-0180.

7. Nooshin Mojahed, Mohammad Ali Mohammadkhani, and Ashraf Mohamadkhani, "Climate Crises and Developing Vector-Borne Diseases: A Narrative Review," *Iranian Journal of Public Health* 51, no. 12 (December 26, 2022): 2664–73, https://www.ncbi.nlm.nih.gov/pmc/articles/PMC9874214/#:~:text=Rising%20temperatures%20favor%20agricultural%20pests,fever%20(2%2C%203.

CHAPTER 4

1. Paul Pastelok, interview by the author, June 5, 2023.

2. Joey Knight, "On Bright, Blissful Day, Bucs Knew How to Shine in 1979 Playoff Debut," *Tampa Bay Times*, January 14, 2022, https://www.tampabay.com/sports/bucs/2022/01/14/on-bright-blissful-day-bucs-knew-how-to-shine-in-1979-playoff-debut/#:~:text=Las%20Vegas'%20line%20favored%20the,Carmichael%2C%20at%20minus%2D4.5.

3. Anthony R. Wood, "Feeling the Effects of Weather," *Philadelphia Inquirer*, February 23, 1988, HS-7.

4. "Local Climatological Summary," National Oceanic and Atmospheric Administration, December 1979, https://www.ncei.noaa.gov/pub/orders/IPS/IPS-7A4A1A6A-3300-499A-9643-57640C0A5572.pdf.

5. Glenn R. McGregor and Jennifer K. Vanos, "Heat: A Primer for Public Health Researchers," *Public Health* 161 (August 2018): 138–46, https://www.sciencedirect.com/science/article/abs/pii/S0033350617303785.

6. Sarah Johnson, interview by the author, June 7, 2024.

7. "Hypothalamus," Cleveland Clinic, March 16, 2022, https://my.clevelandclinic.org/health/articles/22566-hypothalamus.

8. Institute for Quality and Efficiency in Healthcare (IQWiG), "In Brief: How Is Body Temperature Regulated and What Is Fever?," *InformedHealth.org* (Cologne, Germany: IQWiG, 2006), updated December 2022, https://www.ncbi.nlm.nih.gov/books/NBK279457/#:~:text=Our%20internal%20body%20temperature%20is,body%20generates%20and%20maintains%20heat.

9. Chantal A. Vella and Len Kravitz, "Staying Cool When Your Body Is Hot," University of New Mexico, accessed July 10, 2023, https://www.unm.edu/~lkravitz/Article%20folder/thermoregulation.html.

10. "Sweaty Hands," SweatHelp.org, accessed July 12, 2023, https://www.sweathelp.org/where-do-you-sweat/sweaty-hands.html.

11. "Sweaty Hands and Feet," *American Family Physician* 69, no. 5 (2004): 1121, https://www.aafp.org/pubs/afp/issues/2004/0301/p1121.html.

12. Katie McCallum, "How Sweat Works: Why We Sweat When We're Hot, as Well as When We're Not," August 19, 2020, https://www.houstonmethodist.org/blog/articles/2020/aug/how-sweat-works-why-we-sweat-when-we-are-hot-as-well-as-when-we-are-not/#:~:text=As%20soon%20as%20your%20body's,just%20dripping%20off%20of%20you.

13. National Weather Service Mount Holly (@NWS Mount Holly), "We've heard from a lot of people that it seems exceptionally uncomfortable. While the heat and humidity we have presently is nothing unusual, June was," Twitter, July 7, 2023, 6:11 a.m., https://twitter.com/nws_mountholly/status/1677304376229257216?s=42&t=w-zNLuH6T0al_HaFicqwpQ.

14. "Dew Point vs. Humidity," National Weather Service, National Oceanic and Atmospheric Administration, accessed June 11, 2024, https://www.weather.gov/arx/why_dewpoint_vs_humidity#:~:text=The%20dew%20point%20is%20the,water%20in%20the%20gas%20form.

15. John Bannister Tabb, "The Tax-Gatherer," All Poetry, accessed June 11, 2024, https://allpoetry.com/The-Tax-Gatherer.

16. Elizabeth Dougherty, "Why Do We Sweat More in High Humidity?," MIT School of Engineering, October 11, 2011, https://engineering.mit.edu/engage/ask-an-engineer/why-do-we-sweat-more-in-high-humidity/.

17. Anthony R. Wood, "It's Been Way Muggier Than This in Philly, But the Trends Are Disturbing," *Philadelphia Inquirer*, July 28, 2023, https://www.inquirer.com/weather/humidity-philadelphia-weather-heat-dewpoint-20230728.html.

18. Lans P. Rothfusz, "The Heat Index 'Equation' (or, More Than You Ever Wanted to Know about Heat Index)," National Weather Service, accessed August 8, 2023, https://www.weather.gov/media/ffc/ta_htindx.PDF.

19. Anthony R. Wood, "What It Really Means When They Say It Will Feel Like 105 in Philly. It's Probably Not What You Think," *Philadelphia Inquirer*, July 27, 2023, https://www.inquirer.com/weather/excessive-heat-warning-philly-health-emergency-20230727.html.

20. Grahame M. Budd, "Web-Bulb Globe Temperature (WBGT)—Its History and Its Limitations," *Journal of Science and Medicine in Sport* 11, no. 1 (January 2008): 20–32, https://doi.org/10.1016/j.jsams.2007.07.003.

21. "Wet Bulb Globe Temperature," National Weather Service, National Oceanic and Atmospheric Administration, accessed August 8, 2023, https://www.weather.gov/arx/wbgt.

22. Wood, "What It Really Means."

23. Sarah Johnson, email message to author, June 7, 2024.

24. Shayna K. Fever, Jonathan D. W. Kahl, Amy E. Kalkbrenner, Rosa M. Cerón Bretón, and Julia G. Cerón Bretón, "A New Combined Air Quality and Heat Index

in Relation to Mortality in Monterrey, Mexico," *International Journal of Environmental Research and Public Health* 19, no. 6 (2022): 3299, https://doi.org/10.3390/ijerph19063299.

25. S. Solberg, Ø. Hov, A. Søvde, I.S A. Isaksen, P. Coddeville, H. De Backer, C. Forster, Y. Orsolini, and K. Uhse, "European Surface Ozone in the Extreme Summer 2003," *Journal of Geophysical Research* 113, no. 1 (December 10, 2008), https://agupubs.onlinelibrary.wiley.com/doi/epdf/10.1029/2007JD009098.

26. "Ozone," American Lung Association, accessed September 30, 2024, https://www.lung.org/clean-air/outdoors/what-makes-air-unhealthy/ozone.

27. "Surface-Level Ozone," National Aeronautics and Space Administration, accessed June 11, 2024, https://airquality.gsfc.nasa.gov/surface-level-ozone.

28. "Surface-Level Ozone."

29. "Trends in Ozone Adjusted for Weather Conditions," U.S. Environmental Protection Agency, last modified May 6, 2024, https://www.epa.gov/air-trends/trends-ozone-adjusted-weather-conditions#:~:text=The%20adjusted%20trends%20show%20an,in%202022%20compared%20to%202021.

30. "Surface-Level Ozone."

31. "Wildfire Smoke Crosses U.S. on Jet Stream," National Aeronautics and Space Administration, September 5, 2017, https://www.nasa.gov/image-article/wildfire-smoke-crosses-u-s-jet-stream/.

32. Anthony R. Wood, "Smoke from the Pacific Northwest Wildfires Lingers over Philly—And Much of the Nation," *Philadelphia Inquirer*, last modified July 21, 2021, https://www.inquirer.com/weather/wildfire-smoke-philadelphia-air-quality-alert-climate-20210720.html.

33. "Wildland Fires and Smoke," US Environmental Protection Agency, last modified May 21, 2024, https://www.epa.gov/air-quality/wildland-fires-and-smoke.

34. Anthony R. Wood, "Canadian Smoke Is Filling Philly's Skies and Is Expected to Linger through Wednesday," *Philadelphia Inquirer*, last modified June 6, 2023, https://www.inquirer.com/weather/philadelphia-weather-forecast-air-quality-red-flag-dry-thunderstorms-fire-20230606.html.

35. Anthony R. Wood, Rob Tornoe, and Frank Kummer, "The Canadian Smoke over Philly Is Likely Unprecedented, and It May Be Here the Rest of the Week," *Philadelphia Inquirer*, updated June 7, 2023, https://www.inquirer.com/weather/wildfire-smoke-philadelphia-canada-air-quality-alert-forecast-20230607.html.

36. Wood et al., "The Canadian Smoke over Philly."

37. Seth Borenstein, Mary Katherine Wildeman, and Anita Snow, "AP Analysis Finds 2023 Set Record for US Heat Deaths, Killing in Areas That Used to Handle the Heat," *AP News*, May 31, 2024, https://apnews.com/article/record-heat-deadly-climate-change-humidity-south-11de21a526e1cbe7e306c47c2f12438d.

38. "Extreme Heat," Ready.gov, last modified June 11, 2024, https://www.ready.gov/heat.

CHAPTER 5

1. Dr. P. Albert Kratzer, *The Climate of Cities* (Boston: Vieweg and Sohn, 1956), p.11, https://urban-climate.org/wp-content/uploads/2023/03/AlbertKratzer_TheClimate OfCities.pdf.

2. Luke Howard, *The Climate of London*, vol. 1 (London: Harvey and Darton, 1833), p. 2, https://docs.ufpr.br/~feltrim/LIVROS/LukeHoward_Climate-of-London-V1.pdf.

3. Howard, *Climate of London*, p. 7.

4. Howard, *Climate of London*, p. 7.

5. Kratzer, *Climate of Cities*, p. 5.

6. Kratzer, *Climate of Cities*, p. 6.

7. "The Growth of Cities," Digital History, accessed May 28, 2024, https://www .digitalhistory.uh.edu/disp_textbook.cfm?smtID=2&psid=3514.

8. Maria Popova, "How the Clouds Got Their Names and How Goethe Popularized Them with His Science-Inspired Poems," *The Marginalian*, accessed October 17, 2023, https://www.themarginalian.org/2015/07/07/the-invention-of-clouds-luke-howard -hamblyn/.

9. "Luke Howard, the Namer of Clouds," Cloud Appreciation Society, November 29, 2022, https://cloudappreciationsociety.org/luke-howard-the-namer-of-clouds/.

10. "Luke Howard: The Man Who Named the Clouds," Weather People and History, The Weather Doctor, May 1, 1999, http://www.heidorn.info/keith/weather/history/howard .htm.

11. Kratzer, *Climate of Cities*, p. 2.

12. J. Marshall Shepherd, Harold Pierce, and Andrew J. Negri, "Rainfall Modification by Major Urban Areas: Observations from Spaceborne Rain Radar on the TRMM Satellite," *Journal of Applied Meteorology and Climatology* 41, no. 7 (July 1, 2002): 689– 701, https://journals.ametsoc.org/view/journals/apme/41/7/1520-0450_2002_041_0689 _rmbmua_2.0.co_2.xml.

13. Gordon Manley, "On the Frequency of Snowfall in Metropolitan England," *Quarterly Journal of the Royal Meteorological Society* 84, no. 359 (1958): 70–72, https://doi.org /10.1002/qj.49708435910.

14. "The Helmut E. Landsberg Award," Awards & Honors, American Meteorological Society, accessed May 28, 2024, https://www.ametsoc.org/index.cfm/ams/about -ams/ams-awards-honors/awards/awards-for-outstanding-contributions/the-helmut-e -landsberg-award/.

15. Jimmy Stamp, "James W. Rouse's Legacy of Better Living Through Design," *Smithsonian Magazine*, April 23, 2014, https://www.smithsonianmag.com/history/james -w-rouses-legacy-better-living-through-design-180951187/.

16. "History—Explore Columbia," Columbia Association, accessed May 28, 2024, https://columbiaassociation.org/explore-columbia/history/.

17. *Kiddle Encylopedia*, s.v. "Columbia, Maryland Facts for Kids," last modified May 14, 2024, https://kids.kiddle.co/Columbia,_Maryland.

18. Helmut E. Landsberg, "Atmospheric Changes in a Growing Community (the Columbia, Maryland Experience)," *Urban Ecology* 4, no. 1 (May 1979): 55, https://doi .org/10.1016/0304-4009(79)90023-8.

19. Landsberg, "Atmospheric Changes," p. 54.

20. Landsberg, "Atmospheric Changes," p. 78.

21. Landsberg, "Atmospheric Changes," pp. 58–67.

22. "Learn About Heat Islands," Heat Islands, U.S. Environmental Protection Agency, updated August 28, 2023, https://www.epa.gov/heatislands/learn-about-heat-islands#:~: text=A%20review%20of%20research%20studies,2%E2%80%935%C2%B0F%20higher.

23. Kratzer, *Climate of Cities*, p. 2.

24. Frank Kummer and John Duchneskie, "These Philly Neighborhoods Get the Worst of the Summer Heat," *Philadelphia Inquirer*, July 26, 2023, https://www.inquirer .com/science/climate/philadelphia-urban-heat-island-climate-change-climate-central -20230726.html.

25. Karine Laaidi, Abdelkrim Zeghnoun, Bénédicte Dousset, Philippe Bretin, Stépha-nie Vandentorren, Emmanuel Giraudet, and Pascal Beaudeau, "The Impact of Heat Islands on Mortality in Paris during the August 2003 Heat Wave," *Environmental Health Perspectives* 120, no. 2 (September 2011): 254–59, https://ehp.niehs.nih.gov/doi/full/10 .1289/ehp.1103532.

26. S. Vandentorren, P. Bretin, A Zeghnoun, L. Mandereau-Bruno, A. Croisier, C. Cochet, J. Ribéron, I. Siberan, B. Declercq, and M. Ledrans, "August 2003 Heat Wave in France: Risk Factors for Death of Elderly People Living at Home," *European Journal of Public Health* 16, no. 6 (December 2006): 583–91, https://academic.oup.com/eurpub/ article/16/6/583/587693.

27. Jan C. Semenza, Carol H. Rubin, Kenneth H. Falter, Joel D. Selanikio, W. Dana Flanders, Holly L. Howe, and John L. Wilhelm, "Heat-Related Deaths during the July 1995 Heat Wave in Chicago," *New England Journal of Medicine* 335, no. 2 (July 1996): 84–90, https://www.nejm.org/doi/full/10.1056/nejm199607113350203.

28. Jim Angel, "The 1995 Heat Wave in Chicago, Illinois," State Climatologist Office for Illinois, accessed May 28, 2024, https://www.isws.illinois.edu/statecli/general/ 1995chicago.htm#:~:text=The%20heat%20wave%20in%20July,hundreds%20of% 20fatalities%20each%20year.

29. "Basics of Climate Change," Climate Change Science, U.S. Environmental Pro-tection Agency, updated April 2, 2024, https://www.epa.gov/climatechange-science /basics-climate-change#:~:text=Warmer%20air%20holds%20more%20moisture,further %20amplifying%20the%20warming%20effect.&text=For%20more%20information %20on%20greenhouse%20gases%2C%20see%20Greenhouse%20Gas%20Emissions.

30. "Precipitation, Ground Cover, and Temperature," Meteo 3: Introductory Mete-orology, PennState College of Earth and Mineral Sciences Department of Meteorol-ogy and Atmospheric Science, accessed May 28, 2024, https://www.e-education.psu .edu/meteo3/l3_p9.html#:~:text=During%20summer%2C%20when%20dew%20points ,nocturnal%20cooling%20in%20the%20city.

31. U.S. Environmental Protection Agency, "Trees and Vegetation," in *Reducing Urban Heat Islands: Compendium of Strategies Draft* (Washington, DC: EPA, 2008), 3, https:// www.epa.gov/sites/default/files/2017-05/documents/reducing_urban_heat_islands_ch_2 .pdf.

32. Tamara Iungman, Marta Cirach, Federica Marando, Evelise Pereira Barboza, Sasha Khomenko, and Pierre Masselot, "Cooling Cities through Urban Green Infrastructure: A Health Impact Assessment of European Cities," *The Lancet* 401, no. 10376 (February 2023): 577–89, https://www.thelancet.com/journals/lancet/article/PIIS0140-6736%2822%2902585-5/fulltext.

33. Landsberg, "Atmospheric Changes," pp. 53–81.

34. Landsberg, "Atmospheric Changes," p. 81.

CHAPTER 6

1. Jean-Marie Robine, Siu Lan K. Cheung, Sophie Le Roy, Herman Van Oyen, Clare Griffiths, Jean-Pierre Michel, and François Richard Herrmann, "Death Toll Exceeded 70,000 in Europe during the Summer of 2003," *Comptes Rendus Biologies* 331, no. 2 (February 2008): 171–78, https://doi.org/10.1016/j.crvi.2007.12.001.

2. Eric Klinenberg, interview by the author, August 22, 2023.

3. Joan Ballester, Marcos Quijal-Zamorano, Raúl Fernando Méndez Turrubiates, Ferran Pegenaute, François R. Herrmann, Jean Marie Robine, Xavier Basagaña, Cathryn Tonne, Josep M. Antó, and Hicham Achebak, "Heat-Related Mortality in Europe during the Summer of 2022," *Nature Medicine* 29 (July 10, 2023): 1857–66, https://www.nature.com/articles/s41591-023-02419-z.

4. Eric Klinenberg, interview by the author.

5. Anthony R. Wood, "Heat Is a Decidedly Modern Killer," *Philadelphia Inquirer*, September 8, 2003, E5.

6. "Chicago, IL Seasonal Summer Temperature Rankings," National Weather Service, National Oceanic and Atmospheric Administration, accessed June 3, 2024, https://www.weather.gov/lot/Chicago_summer_temps.

7. Eric Klinenberg, "Dying Alone, an interview with Eric Klinenberg, author of *Heat Wave: A Social Autopsy of Disaster in Chicago*," University of Chicago Press, 2002, https://press.uchicago.edu/Misc/Chicago/443213in.html.

8. Klinenberg, interview by the author.

9. Klinenberg, interview by the author.

10. Klinenberg, "Dying Alone."

11. "Local Climatological Data: Monthly Summary," National Centers for Environmental Information, National Oceanic and Atmospheric Administration, July 1993, https://www.ncei.noaa.gov/pub/orders/IPS/IPS-A21AF743-363D-488B-AB7D-F1061A0929A9.pdf.

12. Diana Marder and Walter F. Roche Jr., "19 More Victims of Heat Wave Are Found," *Philadelphia Inquirer*, July 14, 1993, A1.

13. "U.S. Climate Extremes Index (CEI)," National Centers for Environmental Information, National Oceanic and Atmospheric Administration, accessed June 1, 2024, https://www.ncei.noaa.gov/access/monitoring/cei/graph.

14. David L. Cohen, interview by the author, August 25, 2023.

15. Marder and Roche, "19 More Victims."

16. Cohen, interview by the author.

17. Marder and Roche, "19 More Victims."

18. Diana Marder, "U.S. to Check Reports of Heat Deaths in City," *Philadelphia Inquirer*, August 20, 1993, B1.

19. Diana Marder and Anthony R. Wood, "CDC Confirms City's Counting of Heat Deaths," *Philadelphia Inquirer*, June 13, 1994, A1.

20. Ronald H. Brown, James Baker, and Elbert W. Friday Jr., *Natural Disaster Survey Report July 1995 Heat Wave* (Silver Spring, MD: U.S. Department of Commerce, National Oceanic and Atmospheric Administration, and National Weather Service, 1995), https://www.weather.gov/media/publications/assessments/heat95.pdf.

21. Marder and Wood, "CDC Confirms City's Counting."

22. Klinenberg, "Dying Alone."

23. Edmund R. Donoghue, interview by the author, August 28, 2023.

24. Dean Iovino, interview by the author, August 28, 2023.

25. Klinenberg, "Dying Alone."

26. Donoghue, interview by the author.

27. M. J. Fennessy and J. L. Kinter III, "Climatic Feedbacks during the 2003 European Heat Wave," *Journal of Climate* 24, no. 23 (December 1, 2011): 5953–67, https://journals.ametsoc.org/view/journals/clim/24/23/2011jcli3523.1.xml.

28. Jessica Phelan, "Twenty Years after Deadly 2003 Heatwave, What Has France Learned?," RFI (Radio France Internationale), last modified August 8, 2023, https://www.rfi.fr/en/france/20230808-twenty-years-after-the-deadly-2003-heatwave-what-lessons-has-france-learned.

29. Abderrezak Bouchama, "The 2003 European Heat Wave," *Intensive Care Medicine* 30 (January 1, 2004): 1–3, https://link.springer.com/article/10.1007/s00134-003-2062-y 9–11.

30. Phelan, "Twenty Years."

31. Bouchama, "2003 European Heat Wave."

32. Phelan, "Twenty Years."

33. Bouchama, "2003 European Heat Wave."

34. Satbyul Estella Kim, Masahiro Hashizume, Ben Armstrong, Antonio Gasparrini, Kazutaka Oka, Yasuaki Hijioka, Ana M. Vicedo-Cabrera, and Yasushi Honda, "Mortality Risk of Hot Nights: A Nationwide Population-Based Retrospective Study in Japan," *Environmental Health Perspectives* 131, no. 5, https://doi.org/10.1289/EHP11444.

35. "Explore All Countries—France," The World Factbook, CIA.gov, last modified May 29, 2024, https://www.cia.gov/the-world-factbook/countries/france/.

36. Bouchama, "2003 European Heat Wave."

37. "Ground-Level Ozone Basics," U.S. Environmental Protection Agency, accessed June 1, 2024, https://www.epa.gov/ground-level-ozone-pollution/ground-level-ozone-basics#:~:text=air%20emission%20sources.-,How%20does%20ground%2Dlevel%20ozone%20form%3F,volatile%20organic%20compounds%20(VOC).

38. "Report on Behalf of the Commission of Inquiry into the Health and Social Consequences of the Heatwave" (English translation), National Assembly of France, vol. 1 (Feb. 25, 2005), https://www.assemblee-nationale.fr/12/rap-enq/r1455-t1.asp#P549_44499.

39. Bouchama, "2003 European Heat Wave."

40. "Report on Behalf of the Commission of Inquiry into the Health and Social Consequences of the Heatwave."

41. Anthony R. Wood, "It's the Lingering Night Heat that Kills," *Philadelphia Inquirer*, August 14, 2000, C1.

42. "Report on Behalf of the Commission of Inquiry into the Health and Social Consequences of the Heatwave."

43. Phelan, "Twenty Years."

44. Brown, Baker, and Friday, *Natural Disaster Survey Report*.

45. Kristie L. Ebi, Thomas J. Teisberg, Laurence S. Kalkstein, Lawrence Robinson, and Rodney F. Weiher, "Heat Watch/Warning Systems Save Lives: Estimated Costs and Benefits for Philadelphia 1995–98," *Bulletin of the American Meteorological Society* 85, no. 8 (August 1, 2004): 1067–74, https://doi.org/10.1175/BAMS-85-8-1067.

46. Phelan, "Twenty Years."

47. Alan Barreca, Karen Clay, Olivier Deschenes, Michael Greenstone, and Joseph S. Shapiro, "Adapting to Climate Change: The Remarkable Decline in the U.S. Temperature–Mortality Relationship Over the 20th Century," *Journal of Political Economy* 124, no. 1 (January 5, 2016): 105–59, https://eml.berkeley.edu/~saez/course131/barrecaetalJPE16.pdf.

ADDENDUM

1. Rachel Reiff Ellis, "What Air Conditioning Does to Your Body," last modified August 25, 2023, https://www.webmd.com/a-to-z-guides/ss/slideshow-what-ac-does-to-your-body.

2. Gail Cooper, *Air Conditioning America: Engineers and the Controlled Environment, 1900–1960* (Baltimore: Johns Hopkins University Press, 1998), p. 8.

3. John Gladstone, "John Gorrie, the Visionary," *ASHRAE Journal* (December 1998): 29–35, https://www.ashrae.org/file%20library/about/mission%20and%20vision/ashrae%20and%20industry%20history/john-gorrie--the-visionary.pdf.

4. "A Brief History of Malaria," in *Saving Lives, Buying Time: Economics of Malaria Drugs in an Age of Resistance*, ed. Kenneth J. Arrow, Claire B. Panosian, and Hellen Gelband (Washington, DC: National Academies Press, 2004), https://www.ncbi.nlm.nih.gov/books/NBK215638/.

5. George L. Chapel, "Gorrie's Fridge," University of Florida Physics Department, accessed June 12, 2024, http://www.phys.ufl.edu/~ihas/gorrie/fridge.htm.

6. Gladstone, "John Gorrie," p. 31.

7. Chapel, "Gorrie's Fridge."

8. Gladstone, "John Gorrie," p. 31.

9. "10 Tips to Keep You and Your House Cool This Summer," Good Living, January 5, 2024, https://www.environment.sa.gov.au/goodliving/posts/2018/01/keeping-house-cool-efficiently.

10. Ruth E. Mier, "More About Dr. John Gorrie and Refrigeration," *Florida Historical Quarterly* 26, no. 2 (1947): 5, https://stars.library.ucf.edu/cgi/viewcontent.cgi?article=2280&context=fhq.

11. Gladstone, "John Gorrie," p. 31.

12. Anthony R. Wood, "The Coming Storm Might Become a 'Bomb Cyclone.' Just What Does That Mean?," *Philadelphia Inquirer*, January 27, 2022, https://www.inquirer.com/weather/bomb-cyclone-snow-philadelphia-weather-forecast-20220127.html.

13. Gladstone, "John Gorrie," p. 32.

14. Gladstone, "John Gorrie," pp. 32–34.

15. John Gorrie, "Improved Process for the Artificial Production of Ice," US Patent 8080A, issued May 6, 1851, https://patents.google.com/patent/US8080A/en.

16. Minna Scherlinder Morse, "Chilly Reception," *Smithsonian Magazine*, July 2002, https://www.smithsonianmag.com/history/chilly-reception-66099329/.

17. Cooper, *Air Conditioning America*, p. 23.

18. "Buffalo Forge Company," Buffalo Architecture and History, accessed June 15, 2024, https://www.buffaloah.com/h/bfloforge/index.html.

19. "Science: Infant's Father," *Time*, December 3, 1934, https://content.time.com/time/subscriber/article/0,33009,930007,00.html.

20. Cooper, *Air Conditioning America*, p. 23.

21. "The Invention That Changed the World," Carrier, accessed June 15, 2024, https://www.williscarrier.com/weathermakers/1876–1902/.

22. "Judge Magazine," Spartacus Educational, accessed June 15, 2024, https://spartacus-educational.com/ARTjudge.htm.

23. "The Invention That Changed the World."

24. John Blythe, "May 1906: Stuart Cramer and Air Conditioning," *NC Miscellany* (blog), University of North Carolina, May 1, 2012, https://blogs.lib.unc.edu/ncm/2012/05/01/this_month_may_1906/.

25. Blythe, "May 1906."

26. "Frequently Asked Questions," Carrier, accessed June 12, 2024, https://www.williscarrier.com/faqs/.

27. Blythe, "May 1906."

28. "Frequently Asked Questions."

29. Nancy Blumenstalk Mingus, *Buffalo Business Pioneers: Innovation in the Nickel City* (Charleston, SC: The History Press, 2021), p. 66, https://www.google.com/books/edition/Buffalo_Business_Pioneers_Innovation_in/K-IJEAAAQBAJ?hl=en&gbpv=1&dq=%22Carrier+Air+Conditioning+Company+of+America%22+1908&pg=PA66&printsec=frontcover.

30. "The Launch of Carrier Air Conditioning Company," Carrier, accessed June 15, 2024, https://www.williscarrier.com/weathermakers/1903-1914/.

31. "Manufactured Weather," Carrier, accessed June 15, 2024, https://www.williscarrier.com/weathermakers/1915-1922/.

32. Brian Roberts, "Air Conditioning American Movie Theaters 1917–1932," Heritage Group, 2, accessed June 15, 2024, http://www.hevac-heritage.org/movie_theatres_USA/ACM_contents-1/[1-1]INTRO.pdf.

33. "Scientifically Cooled!—The Dawn of Air-Condition Theatres," Theatre Historical Society of America, accessed June 15, 2024, https://historictheatres.org/__trashed/.

34. "Beyond the Factory," Carrier, accessed June 15, 2024, https://www.williscarrier.com/weathermakers/1923-1929/.

35. "Time Machine: Air Conditioning and the Cinema," *Air Conditioning Heating Refrigeration News*, June 12, 2000, https://www.achrnews.com/articles/82653-time-machine-air-conditioning-and-the-cinema.

36. Tim Harford, "How Air Conditioning Changed the World," BBC, June 5, 2017, https://www.bbc.com/news/business-39735802.

37. "The Cool History of Air-Conditioning," NPR, August 6, 2006, https://www.npr.org/transcripts/5621406.

38. "Time Machine."

39. "Weathermakers to the World," Carrier, accessed June 15, 2024, https://www.williscarrier.com/weathermakers/1930-1940/.

40. Frank Trippett, "Great American Cooling Machine," *Time*, August 13, 1979, https://content.time.com/time/subscriber/article/0,33009,948569-2,00.html.

41. Amanda Green, "The Cool History of the Air Conditioner," *Popular Mechanics*, December 31, 2014, https://www.popularmechanics.com/home/a7951/history-of-air-conditioning/.

42. "United States Summary," U.S. Census Bureau, 1960 Census of Housing, accessed June 12, 2024, https://www2.census.gov/library/publications/decennial/1960/housing-volume-1/41962442v1p1ch01.pdf.

43. Green, "The Cool History."

44. "Highlights for Air Conditioning in U.S. Homes by State, 2020," U.S. Energy Information Administration, accessed December 14, 2023, https://www.eia.gov/consumption/residential/data/2020/state/pdf/State%20Air%20Conditioning.pdf.

45. Marc Perry, Brian Mendez-Smith, and Lynda Laughlin, "National Archives Will Soon Release 1950 Census, Offering a Look at a Defining Time in Modern U.S. History," U.S. Census Bureau, March 28, 2022, https://www.census.gov/library/stories/2022/03/1950-census-records-window-to-history.html.

46. Trippett, "Great American Cooling Machine."

47. Renee Obringer, Roshanak Nateghi, Debora Maia-Silva, Sayanti Mukherjee, Vineeth CR, Douglas Brent McRoberts, and Rohini Kumar, "Implications of Increasing Household Air Conditioning Use Across the United States Under a Warming Climate," *Earth's Future* 10, no. 1 (January 2022): e2021EF002434, https://doi.org/10.1029/2021EF002434.

48. "Air Conditioners Fuel the Climate Crisis. Can Nature Help?," UN Environment Programme, June 30, 2023, https://www.unep.org/news-and-stories/story/air-conditioners-fuel-climate-crisis-can-nature-help.

49. Rose M. Mutiso, Morgan D. Bazilian, Jacob Kincer, and Brooke Bowser, "Air Conditioning Should Be a Human Right in the Climate Crisis," *Scientific American*, May 10, 2022, https://www.scientificamerican.com/article/air-conditioning-should-be-a-human-right-in-the-climate-crisis/.

50. Karin Lundgren-Kownacki, Elisabeth Dalholm Hornyanszky, Tuan Anh Chu, Johanna Alkan Olsson, and Per Becker, "Challenges of Using Air Conditioning in an Increasingly Hot Climate," *International Journal of Biometeorology* 62, no. 3 (2018): 401–12, https://doi.org/10.1007/s00484-017-1493-z.

51. Alan Barreca, Karen Clay, Olivier Deschenes, Michael Greenstone, and Joseph S. Shapiro, "Adapting to Climate Change: The Remarkable Decline in the US Temperature–Mortality Relationship Over the Twentieth Century," *Journal of Political Economy* 124, no. 1 (January 5, 2016): 105–59, https://eml.berkeley.edu/~saez/course131/barrecaetalJPE16.pdf.

AUTUMN

1. Kaylee Dusang, "Fall Fronts May Bring Mix of Allergy Triggers," Baylor College of Medicine, October 16, 2019, https://www.bcm.edu/news/fall-fronts-may-bring-mix-allergy-triggers.

2. Anthony R. Wood, "How the Autumn Affects Us," *Philadelphia Inquirer*, September 22, 2003, A-1.

3. Robert H. Shmerling, "Thunderstorm Asthma: Bad Weather, Allergies, and Asthma Attacks," Harvard Health Publishing, Harvard Medical School, June 2, 2022, https://www.health.harvard.edu/blog/thunderstorm-asthma-bad-weather-allergies-and-asthma-attacks-202206222766.

4. "4 Facts You Need to Know about Autumn and Your Health," Genesis Medical Associates, October 3, 2014, https://www.genesismedical.org/blog/4-facts-you-need-to-know-about-autumn-and-your-health.

5. "Your Heart in Winter: Tips to Manage Heart Health in the Cold," Northwestern Medicine, November 2022, https://www.nm.org/healthbeat/healthy-tips/Your-Heart-in-Winter#:~:text=The%20cold%20causes%20blood%20vessels,arteries%20constrict%20in%20the%20cold.

6. Francis K. Davis Jr., "Ulcers and Temperature Changes," *Bulletin of the American Meteorological Society* 39, no. 12 (December 1, 1958): 652–54, https://journals.ametsoc.org/view/journals/bams/39/12/1520-0477-39_12_652.xml?tab_body=pdf.

7. John M. de Castro, "Seasonal Rhythms of Human Nutrient Intake and Meal Pattern," *Physiology & Behavior* 50, no. 1 (July 1991): 243–48.

CHAPTER 7

1. Carmen Nigro, "So, Why Do We Call It Gotham, Anyway?," *New York Public Library* (blog), New York Public Library, January 25, 2011, https://www.nypl.org/blog/2011/01/25/so-why-do-we-call-it-gotham-anyway.

2. Helen Sullivan, "'Pure Magic': Snow Falls on Johannesburg for First Time in 11 Years," *Guardian*, July 10, 2023, https://www.theguardian.com/world/2023/jul/11/pure-magic-snow-falls-on-johannesburg-for-first-time-in-11-years#:~:text=%E2%80%9CIt%20happens%20once%20every%2010,little%20rain%20in%20winter%20months.

3. Norman E. Rosenthal, *Winter Blues*, 4th ed. (New York: Guilford Press, 2013), p. 266.

4. "Local Climatological Data Publication," National Centers for Environmental Information, National Oceanic and Atmospheric Administration, accessed May

30, 2024, https://www.ncdc.noaa.gov/IPS/lcd/lcd.html?_page=1&state=NY&stationID =94728&_target2=Next+%3E.

5. "Climate and Monthly Weather Forecast Johannesburg, South Africa," Weather Atlas, accessed May 30, 2024, https://www.weather-atlas.com/en/south-africa/ johannesburg-climate.

6. Rosenthal, *Winter Blues*, p. 10.

7. "Sunrise and Sunset Times in Johannesburg, City of Johannesburg Metropolitan Municipality, Gauteng, 2001, South Africa," Sunrise Sunset, accessed May 30, 2024, https://sunrise-sunset.org/search?location=johannesburg%2C+south+africa.

8. "My Experience with Seasonal Affective Disorder," *Norman E. Rosenthal* (blog), Norman Rosenthal, M.D., August 26, 2008, https://www.normanrosenthal.com/blog /2008/08/norman-rosenthal-seasonal-affective-disorder/.

9. "June 2024—New York—Sunrise and Sunset Calendar," Sunrise Sunset, accessed May 30, 2024, https://sunrise-sunset.org/us/new-york-ny/2024/6.

10. "Warmest and Coldest Months (Jan–Dec) at Central Park (1869 to Present)," National Weather Service, National Oceanic and Atmospheric Administration, updated October 17, 2023, https://www.weather.gov/media/okx/Climate/CentralPark/ warmcoldmonths.pdf.

11. "Local Climatological Data, Monthly Summary," National Centers for Environmental Information, New York, NY Central Park Observatory, November 1976 (see "Observations at 3-hour intervals"), https://www.ncei.noaa.gov/pub/orders/IPS/IPS -700759A6-204F-4EE7-ACEF-F2F20B752833.pdf.

12. Rosenthal, *Winter Blues*, p. 10.

13. Rosenthal, "Seasonal Affective Disorder."

14. Rosenthal, *Winter Blues*, p. 11.

15. "Rosenthal, Norman," Articles, The History of Modern Biomedicine Research Group, Queen Mary University of London, accessed May 30, 2024, http://www .histmodbiomed.org/article/rosenthal-norman.html.

16. Rosenthal, *Winter Blues*, pp. 11–16.

17. Norman Rosenthal, "Norman Rosenthal's Story of Seasonal Affective Disorder," YouTube video, August 26, 2014, https://www.youtube.com/watch?v=Aub4pHyw7oY.

18. Sandy Rovner, "Healthtalk: Seasons of the Psyche," *Washington Post*, June 11, 1981, https://www.washingtonpost.com/archive/lifestyle/1981/06/12/healthtalk-seasons -of-the-psyche/f744a361-b849-4c1f-925d-c32845a11897/.

19. Rosenthal, "Norman Rosenthal's Story."

20. Rosenthal, *Winter Blues*, p. 16.

21. "The Recent History of Seasonal Affective Disorder (SAD)," Multimedia, The History of Modern Biomedicine Research Group, Queen Mary University of London, accessed May 30, 2024, http://www.histmodbiomed.org/witsem/vol51 .html#:~:text=The%20variation%20of%20mood%20with,Mental%20Health% 20in%20Bethesda%2C%20MD.

22. Norman E. Rosenthal, *Defeating SAD: A Guide to Health and Happiness Through All Seasons* (New York: G&D Media, 2023), p. 12.

23. Steven D. Targum and Norman Rosenthal, "Seasonal Affective Disorder," *Psychiatry* (Edgmont) 5, no. 5 (May 2008): 31–33, https://www.ncbi.nlm.nih.gov/pmc/articles/PMC2686645/.

24. Stuart L. Kurlansik and Annamarie D. Ibay, "Seasonal Affective Disorder," *American Family Physician* 86, no. 11 (December 1, 2012): 1037–41, https://www.aafp.org/pubs/afp/issues/2012/1201/p1037.pdf.

25. "Seasonal Affective Disorder," *American Family Physician* 61, no. 5, 1531–32, https://www.aafp.org/pubs/afp/issues/2000/0301/p1531.html.

26. Phyllis C. Zee, in discussion with the author, February 20, 2024.

27. Rosenthal, *Winter Blues*, p. 21.

28. John M. de Castro, "Seasonal Rhythms of Human Nutrient Intake and Meal Pattern," *Physiology & Behavior* 50, no. 1 (July 1991), 243–48, https://doi.org/10.1016/0031-9384(91)90527-u.

29. Chris Bachman, "Do Bears Really Hibernate?," *NFF* (blog), National Forest Foundation, accessed May 30, 2024, https://www.nationalforests.org/blog/do-bears-really-hibernate.

30. Zee, discussion with the author.

31. Rosenthal, *Winter Blues*, p. 75.

32. "5 Surprising Ways Shorter Days Affect Your Brain [Infographic]: And How You Can Adapt," Healthy Tips, Northwestern Medicine, updated December 2022, https://www.nm.org/healthbeat/healthy-tips/Surprising-Ways-Shorter-Days-Affect-Your-Brain-infographic.

33. Melinda A. Ma and Elizabeth H. Morrison, *Neuroanatomy, Nucleus Suprachiasmatic* (Treasure Island, FL: StatPearls Publishing, 2023), https://www.ncbi.nlm.nih.gov/books/NBK546664/#:~:text=The%20suprachiasmatic%20nucleus%20(SCN)%20is,circadian%20rhythms%20in%20the%20body.

34. Alessandra Porcu, Anna Nilsson, Sathwik Booreddy, Samuel A. Barnes, David K. Welsh, and Davide Dulcis, "Seasonal Changes in Day Length Induce Multisynaptic Neurotransmitter Switching to Regulate Hypothalamic Network Activity and Behavior," *ScienceAdvances* 8, no. 35 (September 2022), https://doi.org/10.1126/sciadv.abn9867.

35. Scott La Fee, "How Changes in Length of Day Change the Brain and Subsequent Behavior," Neuroscience News, accessed May 30, 2024, https://neurosciencenews.com/scn-daylight-dopamine-21357/.

36. "5 Surprising Ways Shorter Days Affect Your Brain."

37. Advertising Council, Sponsor/Advertiser, *Why Do You Think They Call It Dope?*, poster created between 1965 and 1980, Library of Congress Prints and Photographs Division, https://www.loc.gov/item/2015649970/.

38. Dictionary.com, Slang Dictionary, s.v. "Dope," accessed May 30, 2024, https://www.dictionary.com/e/slang/dope/.

39. "Dopamine," Articles, Cleveland Clinic, accessed May 30, 2024, https://my.clevelandclinic.org/health/articles/22581-dopamine.

40. "5 Surprising Ways Shorter Days Affect Your Brain."

41. Stephanie Watson, "Serotonin: The Natural Mood Booster," Harvard Medical School, Harvard Health Publishing, accessed May 30, 2024, https://www.health.harvard.edu/mind-and-mood/serotonin-the-natural-mood-booster.

42. Andrew Chu and Roopma Wadhwa, *Selective Serotonin Reuptake Inhibitors* (Treasure Island, FL: StatPearls Publishing, 2023), https://www.ncbi.nlm.nih.gov/books/NBK554406/#:~:text=Selective%20serotonin%20reuptake%20inhibitors%20(SSRIs,safety%2C%20efficacy%2C%20and%20tolerability.

43. "Melatonin," Articles, Cleveland Clinic, accessed May 30, 3024, https://my.clevelandclinic.org/health/articles/23411-melatonin.

44. "5 Surprising Ways Shorter Days Affect Your Brain."

45. Rosenthal, "Norman Rosenthal's Story."

46. Rosenthal, *Winter Blues*, p. 11.

47. Rosenthal, "Norman Rosenthal's Story."

48. Rosenthal, *Winter Blues*, p. 266.

49. Rosenthal, *Winter Blues*, p. 2.

50. Zee, discussion with the author.

51. Anissa Gabbara, "6 Tips to Beat the Winter Blues," Stories and News, University of Michigan Office of University Development, January 26, 2023, https://giving.umich.edu/um/w/6-tips-to-beat-the-winter-blues#:~:text=Exercise%20is%20critically%20important%20in,or%20hopping%20on%20a%20treadmill.

52. Gabbara, "6 Tips to Beat the Winter Blues."

CHAPTER 8

1. G. Michael Allan and Bruce Arroll, "Prevention and Treatment of the Common Cold: Making Sense of the Evidence," *Canadian Medical Association Journal* 186, no. 3 (2014): 190–99, https://doi.org/10.1503/cmaj.121442.

2. "Facts about the Common Cold," American Lung Association, last modified January 22, 2024, https://www.lung.org/lung-health-diseases/lung-disease-lookup/facts-about-the-common-cold#:~:text=Complications%20of%20a%20Cold,infection%20with%20a%20prolonged%20cough.

3. "Editorial: Recent Studies on 'The Common Cold,'" *Canadian Medical Association Journal* 25, no. 4 (October 1931): 455–56, https://www.ncbi.nlm.nih.gov/pmc/articles/PMC382701/pdf/canmedaj00110-0071.pdf.

4. Ronald Eccles, in discussion with the author, February 20, 2024.

5. Ronald Eccles, "Common Cold," *Frontiers in Allergy* 4 (June 22, 2023): 1224988, https://doi.org/10.3389/falgy.2023.1224988.

6. "Facts about the Common Cold."

7. Eccles, discussion with the author.

8. "Common Cold," Disease & Conditions, Mayo Clinic, May 24, 2023, https://www.mayoclinic.org/diseases-conditions/common-cold/symptoms-causes/syc-20351605.

9. Online Etymology Dictionary, s.v. "Cold," accessed May 30, 2024, https://www.etymonline.com/word/cold.

10. "Common Cold," Conditions and Diseases, Johns Hopkins Medicine, accessed May 30, 2024, https://www.hopkinsmedicine.org/health/conditions-and-diseases/common-cold.

11. Hannah Foster, "The Reason for the Season: Why Flu Strikes in Winter," *Science in the News* (blog), Harvard Kenneth C. Griffin Graduate School of Arts and Sciences, December 1, 2014, https://sitn.hms.harvard.edu/flash/2014/the-reason-for-the-season-why-flu-strikes-in-winter/.

12. David Tyrrell and Michael Fielder, *Cold Wars: The Fight Against the Common Cold* (Oxford: Oxford University Press, 2002), p. 2.

13. Ronald Eccles and Olaf Weber, eds., *Common Cold* (Basel, Switzerland: Birkhäuser, 2009), p. 7, https://www.google.com/books/edition/Common_Cold/rRIdiGE42IEC?hl=en&gbpv=1&pg=PA3&printsec=frontcover.

14. Eccles and Weber, *Common Cold*, p. 3.

15. Eccles and Weber, *Common Cold*, pp. 3–4.

16. Tyrrell and Fielder, *Cold Wars*, p. 7.

17. Nancy Gupton, "Benjamin Franklin and the Kite Experiment," Science and Education, The Franklin Institute, June 12, 2017, https://fi.edu/en/science-and-education/benjamin-franklin/kite-key-experiment.

18. "Fact Sheet: Benjamin Franklin's Inventions," Press Releases, Visit Philadelphia, April 6, 2013, https://www.visitphilly.com/media-center/press-releases/fact-sheet-benjamin-franklin-inventions/.

19. H. Parkins, "In Their Own Words: John Adams and Ben Franklin, Part I," *Pieces of History* (blog), U.S. National Archives, June 20, 2012, https://prologue.blogs.archives.gov/2012/06/20/in-their-own-words-john-adams-and-ben-franklin-part-i/.

20. John Adams, "[Monday September 9, 1776]," in *Diary of John Adams, volume 3*, Adams Papers Digital Edition, Massachusetts Historical Society, accessed May 30, 2024, https://www.masshist.org/publications/adams-papers/view?id=ADMS-01-03-02-0016-0187.

21. Lisa Gensel, "The Medical World of Benjamin Franklin," *Journal of the Royal Society of Medicine* 98, no. 12 (December 2005): 534–38, https://doi.org/10.1177/014107680509801209.

22. Adams, "[Monday September 9, 1776]."

23. Parkins, "In Their Own Words."

24. Jane Chai, "The Forging of an Army," Pennsylvania Center for the Book, accessed May 30, 2024, https://pabook.libraries.psu.edu/literary-cultural-heritage-map-pa/feature-articles/forging-army#:~:text=An%20estimated%203%2C000%20soldiers%20died,troops%20stationed%20at%20Valley%20Forge.

25. Tyrrell and Fielder, *Cold Wars*, p. 19.

26. Tyrrell and Fielder, *Cold Wars*, pp. 12–13.

27. "Robert Koch: One of the Founders of Microbiology," Robert Koch Institute, March 12, 2018, https://www.rki.de/EN/Content/Institute/History/rk_node_en.html.

28. Eccles, "Common Cold."

29. Eccles and Weber, *Common Cold*, p. 16.

30. "Editorial: Recent Studies on 'The Common Cold.'"

31. Caroline Richmond, "David Tyrrell," *British Medical Journal* 330, no. 7505 (June 2005): 1451, https://www.ncbi.nlm.nih.gov/pmc/articles/PMC558394/.

32. Eccles, discussion with the author.

33. Tyrrell and Fielder, *Cold Wars*, p. 50.

34. Richmond, "David Tyrrell."

35. Christopher Andrewes, "The Common Cold: Twenty Years' Work on the Common Cold," *Journal of the Royal Society of Medicine* 59, no. 7 (July 1966): 635–37, https://doi.org/10.1177/003591576605900727.

36. Ivan Oransky, "David Tyrrell," *The Lancet* 365, no. 9477 (June 18, 2005): 2084, https://doi.org/10.1016/S0140-6736(05)66722-0.

37. Tyrrell and Fielder, *Cold Wars*, p. 47.

38. Tyrrell and Fielder, *Cold Wars*, p. 55.

39. Oransky, "David Tyrrell."

40. Tyrrell and Fielder, *Cold Wars*, p. 55.

41. Tyrrell and Fielder, *Cold Wars*, p. 60.

42. Richmond, "David Tyrrell."

43. Andrewes, "The Common Cold," pp. 636–37.

44. Richmond, "David Tyrrell."

45. A. P. Herbert, quoted in Terho Heikkinen and Asko Järvinen, "The Common Cold," *The Lancet* 361, no. 9351 (January 2004): 51–59, https://doi.org/10.1016/S0140-6736(03)12162-9.

46. Andrewes, "The Common Cold," p. 636.

47. Ronald Eccles, in discussion with the author.

48. WebMD Editorial Contributors, "8 Natural Tips to Help Prevent a Cold," WebMD, last updated May 16, 2023, https://www.webmd.com/cold-and-flu/11-tips-prevent-cold-flu.

49. Mayo Clinic Staff, "Cold Remedies: What Works, What Doesn't, What Can't Hurt," Mayo Clinic, June 3, 2022, https://www.mayoclinic.org/diseases-conditions/common-cold/in-depth/cold-remedies/art-20046403.

CHAPTER 9

1. "Famous People with Asthma," Asthma Initiative of Michigan, accessed June 4, 2024, https://getasthmahelp.org/famous-people.aspx.

2. "Asthma Trends and Burden," American Lung Association, accessed June 4, 2024, https://www.lung.org/research/trends-in-lung-disease/asthma-trends-brief/trends-and-burden.

3. Joanne Kavanagh, David J. Jackson, and Brian D. Kent, "Over- and Under-Diagnosis in Asthma," *Breathe* 15, no. 1 (March 2019): e20–e27, https://doi.org/10.1183/20734735.0362-2018.

4. "Air Quality—National Summary," U.S. Environmental Protection Agency, last updated June 4, 2024, https://www.epa.gov/air-trends/air-quality-national-summary#:~:text=EPA%20creates%20air%20quality%20trends,has%20improved%20nationally%20since%201980.

5. "Study: Reducing Human-Caused Air," NOAA Research, May 11, 2022, https://research.noaa.gov/2022/05/11/study-reducing-human-caused-air-pollution-in-north-america-and-europe-brings-surprising-result-more-hurricanes/.

6. Kate King, interview by Anthony R. Wood, February 22, 2024.

7. "What Is Asthma?," Health Topics, National Heart, Lung, and Blood Institute, updated April 17, 2024, https://www.nhlbi.nih.gov/health/asthma.

8. "Asthma," Disease & Conditions, Cleveland Clinic, last modified January 19, 2022, https://my.clevelandclinic.org/health/diseases/6424-asthma.

9. "Asthma," World Health Organization, May 6, 2024, https://www.who.int/news-room/fact-sheets/detail/asthma.

10. "Shared Asthma Stories," American Lung Association, accessed June 4, 2024, https://www.lung.org/lung-health-diseases/lung-disease-lookup/asthma/finding-support/share-your-story/shared-asthma-stories?disease=asthma.

11. Beeathome, "When They Don't Know What They Don't Know," May 23, 2023, https://asthma.net/stories/explaining-asthma-symptoms.

12. Anthony R. Wood, "How the Autumn Affects Us," *Philadelphia Inquirer*, September 22, 2003, A-1.

13. R. E. Dales, I. Schweitzer, J. H. Toogood, M. Drouin, W. Yang, J. Dolovich, and J. Boulet, "Respiratory Infections and the Autumn Increase in Asthma Morbidity," *European Respiratory Journal* 9 (1996): 72–77, https://erj.ersjournals.com/content/erj/9/1/72.full.pdf.

14. David Stukus, "Autumn Asthma Attacks," *700 Children's®—A Blog by Pediatric Experts*, October 2, 2013, https://www.nationwidechildrens.org/family-resources-education/700childrens/2013/10/autumn-asthma-attacks.

15. "Weather Fronts," Center for Science Education, University Corporation for Atmospheric Research, accessed June 3, 2024, https://scied.ucar.edu/learning-zone/how-weather-works/weather-fronts.

16. Robert H. Shmerling, "Thunderstorm Asthma: Bad Weather, Allergies, and Asthma Attacks," Harvard Medical School, Harvard Health Publishing, June 22, 2022, https://www.health.harvard.edu/blog/thunderstorm-asthma-bad-weather-allergies-and-asthma-attacks-202206222766.

17. "North American Monsoon Downburst Winds and Dust Storms," National Weather Service, National Oceanic and Atmospheric Administration, accessed June 4, 2024, https://www.weather.gov/abq/northamericanmonsoon-windsdust#:~:text=In%20the%20photo%20to%20the,in%20less%20than%20a%20minute.

18. "Weather Changes and Asthma," Asthma.com, accessed June 4, 2024, https://www.asthma.com/understanding-asthma/asthma-triggers/weather-changes-and-asthma/.

19. "Ragweed Pollen Allergy," Asthma and Allergy Foundation of America, last modified August 2023, https://aafa.org/allergies/types-of-allergies/pollen-allergy/ragweed-pollen/.

20. "Ragweed Pollen Allergy."

21. Anthony R. Wood, "Allergy Alert: Mold Spores Hit Record," *Philadelphia Inquirer*, September 22, 2016, https://www.inquirer.com/philly/blogs/weather/Allergy-alert-Mold-spores-hit-record.html.

22. "Mold Allergy," Asthma and Allergy Foundation of America, last modified July 2022, https://aafa.org/allergies/types-of-allergies/mold-allergy/.

23. "Are You More Likely to Get Sick When the Seasons Change?" Northern Nevada Medical Center, last modified August 16, 2021, https://www.nnmc.com/about/blog/are -you-more-likely-get-sick-when-seasons-change.

24. "World Language Middle School," *U.S. News & World Report*, accessed June 4, 2024, https://www.usnews.com/education/k12/ohio/world-language-middle-school -412549.

25. "Asthma," Resources, National Association of School Nurses, last modified April 2022, https://www.nasn.org/nasn-resources/resources-by-topic/asthma.

26. King, interview.

27. Andrea K. Greiling, Leslie P. Boss, and Lani S. Wheeler, "A Preliminary Investigation of Asthma Mortality in Schools," *Journal of School Health* 75, no. 8 (October 2005): 286–90, https://doi.org/10.1111/j.1746-1561.2005.00039.x.

28. B. Vernon Clark, "School District Is Sued in Sixth Grader's Asthma Death," *Philadelphia Inquirer*, September 3, 2014, https://www.inquirer.com/philly/education /20140904_Wrongful-death_suit_filed_against_Phila__School_District_in_death_of _sixth-grader.html.

29. "Philadelphia Federation of Teachers President Jerry Jordan on Nursing Shortage, Urgent Course Corrections," Targeted News Service, September 10, 2021, https: //advance.lexis.com/api/document?collection=news&id=urn:contentItem:63K3-18R1 -JC11-150K-00000-00&context=1519360.

30. "CCS Director of Health, Family and Community Services Honored," Columbus City Schools, April 13, 2021, https://www.ccsoh.us/site/Default.aspx ?PageType=3&DomainID=4&PageID=1&ViewID=6446ee88-d30c-497e-9316 -3f8874b3e108&FlexDataID=32217.

31. World Health Organization, "Asthma."

32. King, interview.

33. "Increasing Rates of Allergies and Asthma," American Academy of Allergy, Asthma & Immunology, December 27, 2023, https://www.aaaai.org/tools-for-the -public/conditions-library/allergies/prevalence-of-allergies-and-asthma#:~:text =Some%20experts%20believe%20rising%20rates,in%20reduced%20vitamin %20D%20production.

34. "Asthma: The Hygiene Hypothesis," U.S. Food and Drug Administration, last modified March 23, 2018, https://www.fda.gov/vaccines-blood-biologics/consumers -biologics/asthma-hygiene-hypothesis.

35. D. P. Strachan, "Hay Fever, Hygiene, and Household Size," *British Medical Journal* 299, no. 6710 (November 18, 1989): 1259–60, https://doi.org/10.1136/bmj.299.6710 .1259.

36. Michael R. Perkin and David P. Strachan, "The Hygiene Hypothesis for Allergy— Conception and Evolution," *Frontiers in Allergy* 3 (November 24, 2022), https://doi.org /10.3389/falgy.2022.1051368.

37. Petra I. Pfefferle, Corinna U. Keber, Robert M. Cohen, and Holger Garn, "The Hygiene Hypothesis—Learning from But Not Living in the Past," *Frontiers in Immunology* 12 (March 15, 2021): 635935, https://doi.org/10.3389/fimmu.2021.635935.

38. "Increasing Rates of Allergies and Asthma."

39. Timothy C. Borbet, Miranda B. Pawline, Xiaozhou Zhang, Sergei B. Koralov, Anne Müller, and Martin J. Blaser, "Influence of the Early-Life Gut Microbiota on the Immune Responses to an Inhaled Allergen," *Mucosal Immunology* 15, no. 5 (August 2022): 1000–11, https://doi.org/10.1038/s41385-022-00544-5.

40. Andrew Smith, "Early Exposure to Antibiotics Can Cause Permanent Asthma and Allergies," Rutgers University Office for Research, July 28, 2022, https://research.rutgers.edu/news/early-exposure-antibiotics-can-cause-permanent-asthma-and-allergies.

41. Teeranai Sakulchit and Ran D. Goldman, "Acetaminophen Use and Asthma in Children," *Canadian Family Physician* 63, no. 3 (March 2017): 211–13, https://www.ncbi.nlm.nih.gov/pmc/articles/PMC5349720/.

42. Lara J. Akinbami and Cheryl D. Fryar, "Current Asthma Prevalence by Weight Status among Adults: United States, 2001–2014," National Center for Health Statistics, last modified March 16, 2016, https://www.cdc.gov/nchs/products/databriefs/db239.htm.

43. "Overweight & Obesity Statistics," National Institute of Diabetes and Digestive and Kidney Diseases, last modified September 2021, https://www.niddk.nih.gov/health-information/health-statistics/overweight-obesity#:~:text=%25)%20are%20overweight.-,More%20than%202%20in%205%20adults%20(42.4%25)%20have%20obesity,who%20are%20overweight%20(27.5%25.

44. Stella U. Ogunwole, Megan A. Rabe, Andrew W. Roberts, and Zoe Caplan, "Population Under Age 18 Declined Last Decade," United States Census Bureau, August 12, 2021, https://www.census.gov/library/stories/2021/08/united-states-adult-population-grew-faster-than-nations-total-population-from-2010-to-2020.html#:~:text=Population%20Under%20Age%2018%20Declined%20Last%2 0Decade&text=In%202020%2C%20the%20U.S.%20Census,from%20234.6%20million%20in%20 2010.

45. Ubong Peters, Anne E. Dixon, and Erick Forno, "Obesity and Asthma," *Journal of Allergy and Clinical Immunology* 141, no. 4 (April 2018): 1169–79, https://doi.org/10.1016/j.jaci.2018.02.004.

46. Peters et al., "Obesity and Asthma."

47. Editorial Staff, "Asthma and Weight," American Lung Association, last modified August 30, 2023, https://www.lung.org/blog/the-link-between-asthma-weight#:~:text=Fat%20tissue%20produces%20inflammatory%20substances,in%20a%20healthy%20weight%20range.

48. Hanna Sikorska-Szaflik, Joanna Polomska, and Barbara Sozańska, "The Impact of Dietary Intervention in Obese Children on Asthma Prevention and Control," *Nutrients* 14, no. 20 (October 2022): 4322, https://doi.org/10.3390/nu14204322.

49. "Increasing Rates of Allergies and Asthma."

50. "5 Ways to Keep Your Asthma Under Control," World Health Organization, May 2, 2023, https://www.who.int/news-room/feature-stories/detail/dont-let-asthma-hold -you-back-5-ways-to-make-sure-that-you-are-in-control-of-your-asthma.

ADDENDUM

1. "[Monday September 9, 1776. From the Diary of John Adams]," *Founders Online*, National Archives, accessed June 27, 2024, https://founders.archives.gov /?q=john%20adams%20diary%20Monday%20September%209%2C%201776&s =1111311111&sa=&r=6&sr=.

2. "From Benjamin Franklin to Benjamin Rush, 14 July 1773," *Founders Online*, National Archives, accessed June 27, 2024, https://founders.archives .gov/?q=from%20benjamin%20franklin%20to%20benjamin%20rush%20july% 201773&s=1111311111&sa=&r=8&sr=.

WINTER

1. "Out in the Cold," Harvard Health Publishing, Harvard Medical School, January 1, 2010, https://www.health.harvard.edu/staying-healthy/out-in-the-cold.

2. "Snow Shoveling, Cold Temperatures Combine for Perfect Storm of Heart Health Hazards," American Heart Association, January 11, 2024, https://newsroom.heart.org /news/snow-shoveling-cold-temperatures-combine-for-perfect-storm-of-heart-health -hazards.

3. J. J. Cannell, R. Vieth, J. C. Umhau, M. F. Holick, W. B. Grant, S. Madronich, C. F. Garland, and E. Giovannucci, "Epidemic Influenza and Vitamin D," *Epidemiology & Infection* 134, no. 6 (December 2006): 1129–40, https://pubmed.ncbi.nlm.nih.gov /16959053/.

4. R. E. Hope-Simpson, "The Role of Season in the Epidemiology of Influenza," *Journal of Hygiene* 86, no. 1 (February 1981): 35–47, https://doi.org/10.1017/ s0022172400068728.

5. Paul Crawford, "Editorial Perspective: Cabin Fever—The Impact of Lockdown on Children and Young People," *Child & Adolescent Mental Health* 26, no. 2 (May 2021): 167–68, https://www.ncbi.nlm.nih.gov/pmc/articles/ PMC8250659/#:~:text=In%20Cabin%20Fever%3A%20Surviving%20Lockdown ,the%20spread%20of%20the%20virus.

6. B. M. Bower, *Cabin Fever* (Boston: Little, Brown and Company, 1918), p. 1.

CHAPTER 10

1. Randall Osczevski, "The Basis for the New Wind Chill Chart" (remarks at the 15th Conference on Biometeorology/Aerobiology and the 16th International Congress of Biometeorology, Kansas City, MO, October 29, 2002).

2. "In Brief: How Is Body Temperature Regulated and What Is Fever?," National Library of Medicine, National Center for Biotechnology Information, updated December 6, 2022, https://www.ncbi.nlm.nih.gov/books

/NBK279457/#: ~: text = Our %20internal %20body %20temperature %20is ,body%20generates%20and%20maintains%20heat.

3. "Chills," Health Library, Cleveland Clinic, last modified February 11, 2021, https://my.clevelandclinic.org/health/symptoms/21476-chills.

4. Nicholas Morrissey, "Why Do Your Hands Get Cold?," Columbia Doctors, accessed May 29, 2024, https://www.columbiadoctors.org/news/why-do-your-hands-get-cold#:~:text=Cold%20hands%20generally%20occur%20when,%2C%20brain%2C%20vital%20organs.

5. "Hypothermia," Diseases & Conditions, Mayo Clinic, accessed May 29, 2024, https://www.mayoclinic.org/diseases-conditions/hypothermia/symptoms-causes/syc-20352682.

6. Helen Davidson, "Survivors and Bereaved Seek Answers after 21 Deaths in China Ultramarathon," *Guardian*, May 26, 2021, https://www.theguardian.com/world/2021/may/26/survivors-and-bereaved-seek-answers-after-21-deaths-in-china-ultramarathon.

7. Amy Clark, "Hypothermia in Any Season," *Ultra Running Magazine*, April 21, 2016, https://ultrarunning.com/featured/hypothermia-in-any-season/.

8. Torkjel Tveita and Gary C. Sieck, "Physiological Impact of Hypothermia: The Good, the Bad, and the Ugly," *Physiology* 37, no. 2 (March 2022): 69–87, https://journals.physiology.org/doi/full/10.1152/physiol.00025.2021.

9. Mayo Clinic, "Hypothermia."

10. Mayo Clinic, "Hypothermia."

11. "Hypothermia," Conditions and Diseases, Johns Hopkins Medicine, accessed May 29, 2024, https://www.hopkinsmedicine.org/health/conditions-and-diseases/hypothermia.

12. "Frostbite," Fact Sheets, Yale Medicine, accessed May 29, 2024, https://www.yalemedicine.org/conditions/frostbite#: ~: text = Frostbite %20can %20affect %20not%20only,by%20amputating%20the%20affected%20limb.

13. *Oxford English Dictionary*, s.v. "Frostbite," accessed May 29, 2024, https://www.oed.com/dictionary/frostbite_v.

14. Charles Handford, Pauline Buxton, Katie Russell, Caitlin Imray, Scott E. McIntosh, Luanne Freer, Amalia Cochran, and Christopher Imray, "Frostbite: A Practical Approach to Hospital Management," *Extreme Physiology & Medicine* 3 (April 2014): 7, https://www.ncbi.nlm.nih.gov/pmc/articles/PMC3994495/.

15. "Frostbite," Topics, ScienceDirect, accessed May 29, 2024, https://www.sciencedirect.com/topics/medicine-and-dentistry/frostbite.

16. Jack Beresford, "Mountain Climber Reveals What Frostbitten Fingers Look Like Ahead of Amputation," *Newsweek*, October 21, 2021, https://www.newsweek.com/mountain-climber-reveals-frostbitten-fingers-before-amputation-tiktok-fahad-badar-1641328.

17. Jesse Borke, "Hypothermia," Penn Medicine, last modified November 13, 2021, https://www.pennmedicine.org/for-patients-and-visitors/patient-information/conditions-treated-a-to-z/hypothermia.

18. Borke, "Hypothermia."

19. "Local Climatological Data, Monthly Summary," National Centers for Environmental Information, Indianapolis, IN, January 1994, https://www.ncei.noaa.gov/pub/orders/IPS/IPS-8E308126-D437-454A-9267-78C7298F95F7.

20. Amisha Padnani, "Maurine Bluestein, Who Modernized the Wind Chill Index, Dies at 76," *New York Times*, September 14, 2017, https://www.nytimes.com/2017/09/14/science/maurice-bluestein-who-modernized-the-wind-chill-index-dies-at-76.html.

21. Randall Osczevski and Maurice Bluestein, "The New Wind Chill Equivalent Temperature Chart," *Bulletin of the American Meteorological Society* 86, no. 10 (October 2005): 1453–58, https://doi.org/10.1175/BAMS-86-10-1453.

22. Padnani, "Maurice Bluestein."

23. Osczevski and Bluestein, "New Wind Chill Equivalent," p. 1454.

24. Osczevski and Bluestein, "New Wind Chill Equivalent," p. 1457.

25. "Wind Chill Calculator," Wind Chill Chart, National Weather Service, National Oceanic and Atmospheric Administration, accessed May 29, 2024, https://www.weather.gov/ddc/windchillddc.

26. James A. Ruffner and Frank E. Bair, *The Weather Almanac* (Detroit: Gale Research Co., 1974), p. 102.

27. "Wind Chill Chart," Safety, National Weather Service, National Oceanic and Atmospheric Administration, accessed May 29, 2024, https://www.weather.gov/safety/cold-wind-chill-chart.

28. Jeff Sherry, "Paul Siple—The Father of Wind Chill," Hagen History Center (blog), Hagen History Center, January 22, 2021, https://www.eriehistory.org/blog/paul-siple-the-father-of-wind-chill.

29. Charles F. Passel Antarctic Expedition Collection, 1911–2001, Dolph Briscoe Center for American History, University of Texas, Austin, TX, https://txarchives.org/utcah/finding_aids/01519.xml.

30. Charles F. Passel, *The Antarctic Diary of Charles F. Passel*, ed. T. H. Bauman (Lubbock: Texas Tech University Press, 1994), p. xv.

31. Paul A. Siple and Charles F. Passel, "Measurements of Dry Atmospheric Cooling in Subfreezing Temperatures," *Proceedings of the American Philosophical Society* 89, no. 1 (April 1945): 179–99, https://www.jstor.org/stable/985324.

32. "Wind Chill Chart."

33. Rachel Z. Arndt, "The Ridiculous History of Wind Chill," *Popular Mechanics*, December 12, 2016, https://www.popularmechanics.com/science/environment/a14283/the-ridiculous-history-of-wind-chill/.

34. "WindChill," Amarillo, TX, National Weather Service, National Oceanic and Atmospheric Administration, accessed May 29, 2024, https://www.weather.gov/ama/WindChill.

35. Siple and Passel, "Measurements of Dry Atmospheric Cooling," pp. 196–97.

36. Harvey Landford and Leslie R. Fox, "The Wind Chill Index," *Wilderness & Environmental Medicine* 32, no. 3 (September 2021): 395, https://journals.sagepub.com/doi/pdf/10.1016/j.wem.2021.04.005.

37. Michel B. Ducharme and Dragan Brajkovic, "Guidelines on the Risk and Time to Frostbite during Exposure to Cold Winds," in *Prevention of Cold Injuries*, Meeting

Proceedings, RTO (Neuilly-sur-Seine, France, 2005), pp. 2-1-2-10, https://apps.dtic.mil/sti/pdfs/ADA454525.pdf.

38. Wind Chill Temperature Index, National Weather Service, 2001, https://www.weather.gov/media/grr/brochures/wind-chill-brochure.pdf.

39. Ducharme and Brajkovic, "Guidelines on the Risk and Time to Frostbite," pp. 2–1–2–2.

40. Mayo Clinic, "Frostbite."

41. Ducharme and Brajkovic, "Guidelines on the Risk and Time to Frostbite," p. 2–2.

42. Osczevski and Bluestein, "New Wind Chill Equivalent," pp. 1455–58.

43. Siple and Passel, "Measurements of Dry Atmospheric Cooling," p. 177.

44. "Stay Safe in the Extreme Cold," National Weather Service, National Oceanic and Atmospheric Administration, accessed May 29, 2024, https://www.weather.gov/dlh/extremecold#:~:text=Dress%20For%20The%20Cold%3A&text=Use%20synthetic%20fabrics%20that%20wick,your%20lungs%20from%20extreme%20cold.

CHAPTER 11

1. Your Discovery Science, "The Ice Man: Real Superhumans and the Quest for Future Fantastic," July 27, 2014, YouTube video, 1:39, https://www.youtube.com/watch?v=ueODOr0RhIw.

2. Wouter van Marken Lichtenbelt, "Who Is the Iceman?," *Temperature* (Austin, TX) 4, no. 3 (July 2017): 202–5, https://doi.org/10.1080/23328940.2017.1329001.

3. van Marken Lichtenbelt, "Who Is the Iceman?"

4. Joshua A. Waxenbaum, Vamsi Reddy, and Matthew Varacallo, *Anatomy, Autonomic Nervous System* (Treasure Island, FL: StatPearls Publishing, 2023), National Library of Medicine, National Center for Biotechnology Information, https://www.ncbi.nlm.nih.gov/books/NBK539845/#:~:text=The%20autonomic%20nervous%20system%20is,sympathetic%2C%20parasympathetic%2C%20and%20enteric.

5. Wim Hof and Justin Rosales, *Becoming the Iceman: Pushing Past Perceived Limits*, ed. Justin Rosales and Brooke Robinson (Minneapolis: Mill City Press, 2012), pp. 101–2, https://archive.org/details/becoming-the-iceman-wim-hof/page/n101/mode/2up?view=theater.

6. "Out in the Cold," Staying Healthy, Harvard Medical School, Harvard Health Publishing, last modified January 1, 2010, https://www.health.harvard.edu/staying-healthy/out-in-the-cold.

7. Robert Allan, James Malone, Jill Alexander, Salahuddin Vorajee, Mohammed Ihsan, Warren Gregson, Susan Kwiecien, and Chris Mawhinney, "Cold for Centuries: A Brief History of Cryotherapies to Improve Health, Injury and Post-Exercise Recovery," *European Journal of Applied Physiology* 122, no. 5 (2022): 1153–62, https://www.ncbi.nlm.nih.gov/pmc/articles/PMC9012715/.

8. Express, "Cold Enough for You?: Plungapalooza," *Washington Post*, February 2, 2009, https://www.washingtonpost.com/express/wp/2009/02/03/cold_enough_for_you_plungapalooza/.

9. Didrik Espeland, Lous de Weerd, and James B. Mercer, "Health Effects of Voluntary Exposure to Cold Water—A Continuing Subject of Debate," *International Journal*

of Circumpolar Health 81, no. 1 (September 2022): 2111789, https://doi.org/10.1080 /22423982.2022.2111789.

10. "Adipose Tissue (Body Fat)," Body Systems & Organs, Cleveland Clinic, last modified August 18, 2022, https://my.clevelandclinic.org/health/body/24052-adipose -tissue-body-fat.

11. Nancy Schimelpfenig, "A Cold Plunge Could Help Burn Body Fat and Lower Diabetes Risk, Study Says," Healthline, updated October 23, 2023, https://www.healthline .com/health-news/can-ice-baths-help-you-burn-body-fat-new-research-says-yes#How -cold-exposure-burns-fat,–improves-insulin-resistance.

12. Daniel S. Watson, Brenda J. Shields, and Gary A. Smith, "Snow Shovel-Related Injuries and Medical Emergencies Treated in US EDs, 1990 to 2006," *American Journal of Emergency Medicine* 29, no. 1 (January 2011): 11–17, https://doi.org/10.1016/j.ajem .2009.07.003.

13. Deb Balzer, "Mayo Clinic Minute: Heart Health and Dangers of Shoveling Snow," Mayo Clinic, accessed May 29, 2024, https://newsnetwork.mayoclinic.org/discussion/ mayo-clinic-minute-heart-health-and-dangers-of-shoveling-snow/.

14. Aaron Lee, "Safe Snow Shoveling: The Health Benefits and How to Prevent Injuries," *Loyola Medicine's Blog* (blog), Loyola Medicine, November 28, 2023, https://www .loyolamedicine.org/about-us/blog/snow-shoveling-health-benefits-injury-prevention.

15. Zhang Feng, Yang Hu, Sen Yu, Haomiao Bai, Yubo Sun, Weilu Gao, Jia Li, Xiangyang Qin, and·Xing Zhan, "Exercise in Cold: Friend than Foe to Cardiovascular Health," *Life Sciences* 328 (September 2023): 1219223, https://doi.org/10.1016/j.lfs.2023.121923.

16. Casey McIlvaine, "The Benefits of Exercising in Cold Weather," National Personal Training Institute, January 22, 2015, https://nationalpti.org/benefits-exercising-cold -weather/.

17. "How Brown Fat Improves Metabolism," NIH Research Matters, National Institutes of Health, September 10, 2019, https://www.nih.gov/news-events/nih-research -matters/how-brown-fat-improves-metabolism.

18. "Brown Fat," Body Systems & Organs, Cleveland Clinic, accessed May 29, 2024, https://my.clevelandclinic.org/health/body/24015-brown-fat.

19. Tim Newman, "What Are Mitochondria?," Medical News Today, updated June 14, 2023, https://www.medicalnewstoday.com/articles/320875.

20. "The Wonders of Winter Workouts," Staying Healthy, Harvard Medical School, Harvard Health Publishing, December 1, 2018, https://www.health.harvard.edu/staying -healthy/the-wonders-of-winter-workouts.

21. John A. Billon, "Surprising Health Benefits of Cold Weather," Maryland Primary Care Physicians, accessed May 29, 2024, https://www.mpcp.com/articles/healthy -lifestyle/surprising-health-benefits-cold-weather/.

22. "Fighting the World's Deadliest Animal," Global Health, August 14, 2024, https: //www.cdc.gov/global-health/impact/fighting-the-worlds-deadliest-animal.html?CDC _AAref_Val=https://www.cdc.gov/globalhealth/stories/2019/world-deadliest-animal .html.

23. "Where Do Mosquitoes Go in the Winter?," Mosquito Information, Central Mass. Mosquito Control Project, accessed May 29, 2024, https://

www.cmmcp.org/mosquito-information/faq/where-do-mosquitoes-go
-winter#:~:text=In%20temperate%20climates%2C%20adult%20mosquitoes
,of%20females%20in%20late%20summer.

24. Paul Crawford, "Editorial Perspective: Cabin Fever—The Impact of Lockdown on Children and Young People," *Child and Adolescent Mental Health* 26, no. 2 (May 2021): 167–68, https://doi.org/10.1111/camh.12458.

25. Crawford, "Cabin Fever," pp. 167–68.

26. "Daily Insolation Parameters," National Aeronautics and Space Administration, accessed May 29, 2024, https://data.giss.nasa.gov/cgi-bin/ar5/srlocat.cgi.

27. "'Dress Like a Cabbage': Surviving Yakutsk, the World's Coldest City," Reuters, updated January 16, 2023, https://www.reuters.com/world/europe/dress-like-cabbage
-surviving-worlds-coldest-city-2023-01-15/.

28. Marlene Laruelle and Sophie Hohmann, "Yakutsk and Mirnyi Fieldwork Report," *Promoting Urban Sustainability in the Arctic* (blog), George Washington University, August 5, 2017, https://blogs.gwu.edu/arcticpire/tag/yakutsk/.

29. "Manufacturing Companies in Yakutsk, Russian Federation," Dun & Bradstreet, accessed May 29, 2024, https://www.dnb.com/business-directory/company-information
.manufacturing.ru.na.yakutsk.html.

30. Ksenia Acquaviva, "Reflections on Field Work in Yakutsk," *Promoting Urban Sustainability in the Arctic* (blog), George Washington University, June 30, 2017, https://blogs
.gwu.edu/arcticpire/2017/06/30/reflections-on-fieldwork-in-yakutsk/.

31. "Permafrost and Periglacial Studies," Geological and Geophysical Surveys, Department of Natural Resources, accessed May 29, 2024, https://dggs.alaska.gov/hsg/
permafrost.html.

32. Acquaviva, "Reflections on Field Work in Yakutsk."

33. O. V. Tatarinova, E. S. Kylbanova, V. N. Neustroeva, A. N. Semenova, and I. P. Nikitin, "Phenomenon of Super-Longevity in Yakutia," *Advances in Gerontology* 21, no. 2 (2008): PMID 18942361, 198–203, https://pubmed.ncbi.nlm.nih.gov/18942361/.

34. Gerontology Wiki, s.v. "Varvara Semennikova," accessed May 29, 2024, https://
gerontology.fandom.com/wiki/Varvara_Semennikova.

35. Tatarinova et al., "Phenomenon of Super-Longevity in Yakutia."

36. "Life Expectancy at Birth: FE: Republic of Sakha (Yakutia)," Federal State Statistics Service, CEIC, accessed May 29, 2024, https://www.ceicdata.com/en/
russia/life-expectancy-at-birth-by-region/life-expectancy-at-birth-fe-republic-of
-sakha-yakutia#:~:text=State%20Statistics%20Service-kha%20(Yakutia,of%2069
.980%20Year%20for%202021.

37. Susan Casey, "How Iceman Wim Hof Discovered the Secrets to Our Health," *Outside*, updated May 11, 2022, https://www.outsideonline.com/outdoor-adventure/
exploration-survival/wim-hof-method/.

38. van Marken Lichtenbelt, "Who Is the Iceman?" pp. 202–5.

39. Linda Geddes, "Wim Hof Breathing and Cold-Exposure Method May Have Benefits, Study Finds," *Guardian*, March 13, 2024, https://www.theguardian.com/science
/2024/mar/13/wim-hof-breathing-cold-exposure-method-benefits-study.

40. L. T. Buck, I. De Groote, Y. Hamada, B. R. Hassett, T. Ito, and J. T. Stock, "Evidence of Different Climatic Adaptation Strategies in Humans and Non-Human Primates," *Scientific Reports* 9 (July 2019): 11025, https://doi.org/10.1038/s41598-019-47202-8.

41. Laura Buck and Kyoko Yamaguchi, "How Ancient Human Came to Cope with the Cold," *Sapiens*, February 7, 2023, https://www.sapiens.org/biology/humans-cold-environment-adaptations/.

42. Hein A. M. Daanen and Wouter D. van Marken Lichtenbelt, "Human Whole Body Cold Adaptation," *Temperature* (Austin, TX) 3, no. 1 (February 2016): 104–18, https://doi.org/10.1080/23328940.2015.1135688.

43. van Marken Lichtenbelt, "Who Is the Iceman?" pp. 202–5.

44. K. Aleisha Fetters, "13 Winter Workout Tips for Exercising Outdoors No Matter the Weather," Everyday Health, updated January 29, 2024, https://www.everydayhealth.com/healthy-living/fitness/easy-winter-exercise-tips-help-you-stay-fit/.

CHAPTER 12

1. "Extra, Extra! Read All About It: General Mills Doubles Vitamin D in Big G Cereals," General Mills, July 19, 2023, https://www.generalmills.com/news/press-releases/vitamind.

2. "Vitamin D," Dietary Supplement Fact Sheets, last modified November 8, 2022, https://ods.od.nih.gov/factsheets/VitaminD-Consumer/.

3. "Vitamin D Deficiency," Cleveland Clinic, last modified August 2, 2022, https://my.clevelandclinic.org/health/diseases/15050-vitamin-d-vitamin-d-deficiency.

4. Philip Smith, "Michael F. Holick, PhD, MD: The Pioneer of Vitamin D Research," *Life Extension*, August 2023, https://www.lifeextension.com/magazine/2010/9/michael-holick-the-pioneer-of-vitamin-d-research.

5. "Robert D. Ashley, MD," Providers, UCLA Health, accessed June 6, 2024, https://www.uclahealth.org/providers/robert-ashley.

6. "How to Get Enough Vitamin D in the Winter," TUFTS Health Plan, accessed June 6, 2024, https://www.tuftsmedicarepreferred.org/healthy-living/how-get-enough-vitamin-d-winter#:~:text=In%20the%20wintertime%20daylight%20saving,week%20can%20make%20a%20difference.

7. Wedad Z. Mostafa and Rehab A. Hegazy, "Vitamin D and the Skin: Focus on a Complex Relationship: A Review," *Journal of Advanced Research* 6, no. 6 (November 2015): 795, https://www.ncbi.nlm.nih.gov/pmc/articles/PMC4642156/pdf/main.pdf.

8. Mostafa and Hegazy, "Vitamin D and the Skin," 795.

9. Mayo Clinic Staff, "Rickets," Mayo Clinic, February 25, 2021, https://www.mayoclinic.org/diseases-conditions/rickets/symptoms-causes/syc-20351943.

10. "Vitamin D and Your Health: Breaking Old Rules, Raising New Hopes," Harvard Health Publishing, Harvard Medical School, September 13, 2021, https://www.health.harvard.edu/staying-healthy/vitamin-d-and-your-health-breaking-old-rules-raising-new-hopes.

11. "Vitamin D," The Nutrition Source, Harvard T. H. Chan School of Public Health, last modified March 2023, https://www.hsph.harvard.edu/nutritionsource/

vitamin-d/#:~:text=Vitamin%20D%20production%20in%20the,inside%20much%20of%
20the%20time.

12. Sneha Baxi Srivastava, "Vitamin D: Do We Need More Than Sunshine?" *American Journal of Lifestyle Medicine* 15, no. 4 (April 3, 2021): 397–401, https://doi.org/10.1177 /15598276211005689.

13. Ryan Raman, "How to Safely Get Vitamin D from Sunlight," Healthline, last modified April 4, 2023, https://www.healthline.com/nutrition/vitamin-d-from-sun#overview.

14. Julie Corliss, "How It's Made: Cholesterol Production in Your Body," Harvard Health Publishing, Harvard Medical School, February 6, 2017, https://www.health .harvard.edu/heart-health/how-its-made-cholesterol-production-in-your-body.

15. John Bilezikian, "What's the Deal with Vitamin D?," Columbia University Irving Medical Center, August 24, 2022, https://www.cuimc.columbia.edu/news/whats-deal -vitamin-d.

16. James MacDonald, "How Does the Body Make Vitamin D from Sunlight?" JSTOR Daily, July 18, 2019, https://daily.jstor.org/how-does-the-body-make-vitamin-d -from-sunlight/.

17. M. F. Holick, J. A. MacLaughlin, M. B. Clark, S. A. Holick, J. T. Potts, R. R. Anderson, I. H. Blank, J. A. Parrish, and P. Elias, "Photosynthesis of Previtamin D3 in Human Skin and the Physiologic Consequences," *Science* 210, no. 4466 (October 10, 1980): 203–5, https://doi.org/10.1126/science.6251551.

18. Smith, "Michael F. Holick."

19. Liz Szabo, "The Man Who Sold America on Vitamin D—and Profited in the Process," KFF Health News, August 20, 2018, https://kffhealthnews.org/news/how-michael -holick-sold-america-on-vitamin-d-and-profited/.

20. "Calcium and Vitamin D," Bone Health and Osteoporosis Foundation, last modified May 23, 2023, https://www.bonehealthandosteoporosis.org/patients/treatment/ calciumvitamin-d/.

21. "Calcium, Vitamin D, and Your Bones," Mount Sinai Health System, accessed June 6, 2024, https://www.mountsinai.org/health-library/selfcare-instructions/calcium -vitamin-d-and-your-bones.

22. "Calcium and Vitamin D: Important for Bone Health," National Institute of Arthritis and Musculoskeletal and Skin Diseases, accessed June 6, 2024, https://www .niams.nih.gov/health-topics/calcium-and-vitamin-d-important-bone-health.

23. Fernando de la Guía-Galipienso, María Martínez-Ferrand, Néstor Vallecilloe, Carl J. Lavief, Fabian Sanchis-Gomar, and Helios Pareja-Galeano, "Vitamin D and Cardiovascular Health," *Clinical Nutrition* 40, no. 5 (May 2021): 2946–57, https://doi.org/10 .1016/j.clnu.2020.12.025.

24. Smith, "Michael F. Holick."

25. Monique Tello, "Vitamin D: What's the 'Right' Level?" *Harvard Health Blog*, Harvard Health Publishing, Harvard Medical School, April 16, 2020, https://www.health .harvard.edu/blog/vitamin-d-whats-right-level-2016121910893.

26. American Heart Association News, "Vitamin D Is Good for the Bones, but What about the Heart?," American Heart Association, September 17, 2019, https://www.heart .org/en/news/2019/09/17/vitamin-d-is-good-for-the-bones-but-what-about-the-heart.

27. "Vitamin D for Heart Health: Where the Benefits Begin and End," National Heart, Lung, and Blood Institute, September 27, 2022, https://www.nhlbi.nih.gov/news/2022/vitamin-d-heart-health-where-benefits-begin-and-end.

28. Luka Vranić, Ivana Mikolašević, and Sandra Milić, "Vitamin D Deficiency: Consequence or Cause of Obesity?," *Medicina* (Kaunas) 55, no. 9 (August 28, 2019): 541, https://doi.org/10.3390/medicina55090541.

29. Alina Petre, "Can Vitamin D Deficiency Cause Weight Gain?," Healthline, January 7, 2021, https://www.healthline.com/nutrition/low-vitamin-d-and-weight-gain#vitamin-d-weight-gain.

30. Smith, "Michael F. Holick."

31. Caroline Apovian, "Does Vitamin D Deficiency Cause Obesity or Vice Versa?" Medscape, December 27, 2022, https://www.medscape.com/viewarticle/985973.

32. Steven R. Cummings and Clifford Rosen, "VITAL Findings—A Decisive Verdict on Vitamin D Supplementation," *New England Journal of Medicine* 387, no. 4 (July 27, 2022): 368–70, https://doi.org/10.1056/NEJMe2205993.

33. Hope Cristol, "Myths and Facts about Cancer and Vitamin D," last modified September 21, 2023, https://www.webmd.com/cancer/cancer-vitamin-d-myths-facts.

34. "Vitamin D and Cancer," National Cancer Institute at the National Institutes of Health, last modified May 9, 2023, https://www.cancer.gov/about-cancer/causes-prevention/risk/diet/vitamin-d-fact-sheet#:~:text=Higher%20vitamin%20D%20levels%20have,cancers%20(14%E2%80%9317).

35. Theresa Sullivan Barger, "What Do Scientists Know about Vitamin D and Cancer?" *Discover Magazine*, February 20, 2024, https://www.discovermagazine.com/health/what-do-scientists-know-about-vitamin-d-and-cancer.

36. Smith, "Michael F. Holick."

37. Paulette D. Chandler, Wendy Y. Chen, Oluremi N. Ajala, Aditi Hazra, Nancy Cook, Vadim Bubes, I-Min Lee, Edward L. Giovannucci, Walter Willett, Julie E. Buring, and JoAnn E. Manson, "Effect of Vitamin D3 Supplements on Development of Advanced Cancer: A Secondary Analysis of the VITAL Randomized Clinical Trial," *JAMA Network Open* 3, no. 11 (November 18, 2020): e2025850, https://doi.org/10.1001/jamanetworkopen.2020.25850.

38. "Vitamin D Supplements Linked to Lower Risk of Advanced Cancer," Harvard Health Publishing, Harvard Medical School, April 7, 2023, https://www.health.harvard.edu/staying-healthy/vitamin-d-supplements-linked-to-lower-risk-of-advanced-cancer.

39. Sullivan Barger, "What Do Scientists Know."

40. Denis Pereira Gray, "Robert Edgar Hope-Simpson," *BMJ* 327, no. 7423 (November 8, 2003): 1111, https://www.ncbi.nlm.nih.gov/pmc/articles/PMC261759/.

41. R. E. Hope-Simpson, "Sunspots and Flu: A Correlation," *Nature* 275, no. 86 (1978), https://doi.org/10.1038/275086a0.

42. R. E. Hope-Simpson, "The Role of Season in the Epidemiology of Influenza," *Journal of Hygiene* 86, no. 1 (February 1981): 35–47, https://doi.org/10.1017/s0022172400068728.

43. J. J. Cannell, R. Vieth, J. C. Umhau, M. F. Holick, W. B. Grant, S. Madronich, C. F. Garland, and E. Giovannucci, "Epidemic Influenza and Vitamin D,"

Epidemiology and Infection 134, no. 6 (December 2006): 1129–40, https://doi.org/10.1017/S0950268806007175.

44. Cannell et al., "Epidemic Influenza and Vitamin D."

45. David J. G. Slusky and Richard J. Zeckhauser, "Sunlight and Protection against Influenza" (M–RCBG Faculty Working Paper Series, Harvard Kennedy School, Mossavar–Rahmani Center for Business and Government, Cambridge, MA, 2020), https://www.hks.harvard.edu/sites/default/files/centers/mrcbg/files/FWP_2020-03.pdf.

46. Srivastava, "Vitamin D: Do We Need More Than Sunshine?"

47. Matthias Wacker and Michael F. Holick, "Sunlight and Vitamin D," *Dermato-Endocrinology* 5, no. 1 (January 1, 2013): 51–108, https://doi.org/10.4161/derm.24494.

48. "Vitamin D," Dietary Supplement Fact Sheets, National Institutes of Health Office of Dietary Supplements, last modified November 8, 2022, https://ods.od.nih.gov/factsheets/VitaminD-Consumer/.

49. Smith, "Michael F. Holick."

50. Apovian, "Does Vitamin D Deficiency."

51. Colleen Moriarty, "Vitamin D Myths 'D'-bunked," Yale Medicine, March 15, 2018, https://www.yalemedicine.org/news/vitamin-d-myths-debunked.

52. Homa Warren, "Getting Adequate Vitamin D in the Fall and Winter," Baylor College of Medicine, November 9, 2023, https://www.bcm.edu/news/getting-adequate-vitamin-d-in-the-fall-and-winter.

ADDENDUM

1. Edward Bellamy, "The Cold Snap," 1898, Project Gutenberg, accessed June 27, 2024, https://www.gutenberg.org/files/22715/22715-h/22715-h.htm.

INDEX

www.ingramcontent.com/pod-product-compliance
Lightning Source LLC
Chambersburg PA
CBHW020025170225
21802CB00005B/5